T0349832

# Langevin and Fokker-Planck Equations and their Generalizations

*Descriptions and Solutions*

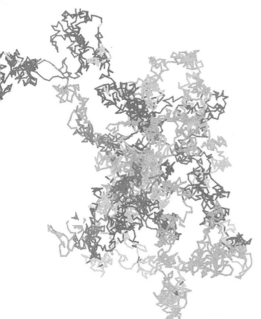

# Langevin and Fokker-Planck Equations and their Generalizations

*Descriptions and Solutions*

Sau Fa Kwok

State University of Maringá, Brazil

 **World Scientific**

NEW JERSEY · LONDON · SINGAPORE · BEIJING · SHANGHAI · HONG KONG · TAIPEI · CHENNAI · TOKYO

*Published by*

World Scientific Publishing Co. Pte. Ltd.
5 Toh Tuck Link, Singapore 596224
*USA office:* 27 Warren Street, Suite 401-402, Hackensack, NJ 07601
*UK office:* 57 Shelton Street, Covent Garden, London WC2H 9HE

**Library of Congress Cataloging-in-Publication Data**
Names: Kwok, Sau Fa, author.
Title: Langevin and Fokker-Planck equations and their generalizations :
 descriptions and solutions / by Sau Fa Kwok (State University of Maringá, Brazil).
Description: New Jersey : World Scientific, 2017. | Includes bibliographical references and index.
Identifiers: LCCN 2017042757 | ISBN 9789813228405 (hardcover : alk. paper)
Subjects: LCSH: Statistical mechanics. | Differential equations, Partial. | Statistical physics.
Classification: LCC QC174.8 .K96 2017 | DDC 530.13--dc23
LC record available at https://lccn.loc.gov/2017042757

**British Library Cataloguing-in-Publication Data**
A catalogue record for this book is available from the British Library.

For any available supplementary material, please visit
http://www.worldscientific.com/worldscibooks/10.1142/9745#t=suppl

Printed in Singapore

This book is dedicated to the memory of my grandfather who illuminated my way, bringing inspiration for my life journey.

# Preface

This book deals with some mathematical formulations of the nonequilibrium statistical physics: The Langevin equation, the Fokker-Planck equation, the continuous time random walk model and their generalizations. Despite the origin, they have been employed to describe nonequilibrium systems in different areas, such as physics, biology, chemistry, hydrology and economics.

The first part of the book complements the classical book on the Langevin and Fokker-Planck equations (H. Risken, The Fokker-Planck equation, 1996). Due to the growing interest in the researches on the generalized Langevin equations, several of them are presented; they are described with some details. The last part is devoted to the continuous time random walk model (CTRW). The book is sketched as follows. Chapter 2 is dedicated to the derivation of the Fokker-Planck equation (for one dimension) from the Langevin equation for different orders of prescription in discretization rules for the stochastic integrals due to the multiplicative white noise. Chapter 3 deals with various methods for solving the Fokker-Planck equation which are not described in the Risken's book, such as the method of similarity solution, the method of characteristics, transformations of diffusion processes into the Wiener process for different orders of prescription in discretization rules for the stochastic integrals, and colored noise. Connection between the Langevin equation and Tsallis distribution is also discussed. Chapter 4 describes the Fokker-Planck equation for several variables which includes the relativistic Brownian motion. Chapter 5 introduces and discusses some generalized Langevin equations which contain non-local operators in time. Chapters 6 and 7 deal with the continuous time random walk model and the derivation of integro-differential Fokker-Planck equation; derivation of an integro-differential Klein-Kramers equation from the generalized Chapman-Kolmogorov equation; methods of solutions; generalization of the integro-differential Klein-Kramers equation for the description of superdiffusive regime.

Finally, I would like to thank Dr. Swee Cheng Lim and Dr. Ke Gang Wang for their suggestions and co-operation.

For comments and suggestions, please, send to the email: kwok@dfi.uem.br.

Maringá, May, 2017                                              *Kwok Sau Fa*

# Contents

# Chapter 1

# Introduction

Diffusion is a very common process in nature and it can be observed in many different systems. In particular, anomalous diffusion processes have been observed in most systems investigated such as bacterial cytoplasm motion [1], conformational fluctuations within a single protein molecule [2,3], fluorescence intermittency in single enzymes [4], the cell migration of two migrating transformed renal epithelial MadinDarby canine kidney (MDCK-F) cell strains [5], the internal protein dynamics for the backbone atoms of hydrated elastin [6] and white and gray matters of a fixed rat brain [7]. Their properties have also been extensively investigated by various approaches in order to model different kinds of probability distributions such as long-range spatial or temporal correlations [8–14]. For instance, we may cite the Langevin equation [15–18], the Fokker-Planck equation [15–17], the continuous-time random walk model (CTRW) [8,19], the generalized Langevin equation [20–22], the fractional Fokker-Planck equation [8] and the nonlinear Fokker-Planck equation [23]. These approaches have also been used to describe numerous systems in various contexts, such as economics, physics, hydrology, chemistry and biology. The diffusion process is classified according to the mean square displacement (MSD)

$$\langle x^2(t) \rangle \sim t^\alpha. \tag{1.1}$$

In the case of normal diffusion, the MSD grows linearly with time ($\alpha = 1$). For $0 < \alpha < 1$ the process is called subdiffusive, and for $\alpha > 1$ the process is called superdiffusive. For $\alpha = 2$ it is referred to as ballistic motion. The well-established property of the normal diffusion described by the Gaussian distribution (the probability distribution is denoted by $\rho(x,t)$, and $\rho(x,t)dx$ is the probability for finding a particle in a position between $x$ and $x + dx$

at time $t$) can be obtained from the ordinary Langevin equation

$$\frac{d\xi}{dt} = \sqrt{D}L(t),$$        (1.2)

where $\xi$ is a stochastic variable, $D$ is the diffusion coefficient and $L(t)$ is the Langevin force which is assumed to be a Gaussian random variable, or from the Fokker-Planck equation

$$\frac{\partial\rho(x,t)}{\partial t} = -\frac{\partial}{\partial x}\left[D_1(x,t)\rho(x,t)\right] + \frac{\partial^2}{\partial x^2}\left[D_2(x,t)\rho(x,t)\right]$$        (1.3)

with vanishing drift coefficient $D_1(x,t) = 0$ and constant diffusion coefficient $D_2(x,t) = D$ [15, 16]; it can also be obtained from an integro-differential diffusion equation

$$\frac{\partial\rho(x,t)}{\partial t} - \int_0^t dt_1 g\left(t-t_1\right)\frac{\partial\rho(x,t_1)}{\partial t_1} = C\frac{\partial}{\partial t}\int_0^t dt_1 g\left(t-t_1\right)\frac{\partial^2\rho(x,t_1)}{\partial x^2}$$        (1.4)

with the exponential function for the waiting time probability distribution, $g(t) = be^{-bt}$ [24]. Anomalous diffusion regimes can also be obtained from the ordinary Fokker-Planck equation, however, they arise from a variable diffusion coefficient which may depend on time and/or space. Besides, in the view of the Langevin approach it may be associated with a multiplicative noise term. In the case of generalized Langevin equation [20, 21], it is described by

$$\frac{dv}{dt} + \int_0^t dt_1 \gamma\left(t-t_1\right)v\left(t_1\right) = F(x) + Q(t)$$        (1.5)

with a unitary mass (where $F(x)$ and $Q(t)$ are deterministic and stochastic forces, respectively).

In other approaches, such as the fractional Langevin equations and generalized Fokker-Planck equations (fractional and nonlinear) [23,25–28], they can also describe anomalous diffusion processes.

This book provides a broad introduction to a rapidly growing area of nonequilibrium statistical physics. The first part of the book complements the classical book on the Langevin and Fokker-Planck equations [15]. Some topics and methods of solutions are presented and discussed in details which are not described in Ref. [15], such as the method of similarity solution, the method of characteristics, transformations of diffusion processes into the Wiener process for different orders of prescription in discretization rules for the stochastic integrals, and harmonic noise.

Due to the growing interest in the research on the generalized Langevin equations, several of them are presented. They are described with some details.

The last part is devoted to the continuous time random walk model (CTRW). The CTRW model is based on the length of a given jump associated with the waiting time elapsing between two successive jumps, and these quantities are connected by a probability density function (PDF) of jumps. Recent research on the integro-differential Fokker-Planck equation derived from the continuous time random walk model is also presented and discussed in details. The equation is worked analytically for linear force and generic waiting time PDF. Generalizations of the integro-differential Fokker-Planck equation are also presented and discussed.

## Bibliography

[1] I. Golding and E. C. Cox, *Phys. Rev. Lett.* **96**, 098102 (2006).

[2] S. C. Kou and X. S. Xie, *Phys. Rev. Lett.* **93**, 180603 (2004).

[3] W. Min, G. B. Luo, B. J. Cherayil, S. C. Kou and X. S. Xie, *Phys. Rev. Lett.* **94**, 198302 (2005).

[4] S. Chaudhury, S. C. Kou and B. J. Cherayil, *J. Phys. Chem. B* **111**, 2377 (2007).

[5] P. Dieterich, R. Klages , R. Preuss and A. Schwab, *Proc. Nat. Acad. Sci. USA* **105**, 459 (2008).

[6] K. Kämpf, F. Klameth, and M. Vogel, *J. Chem. Phys.* **137**, 205105 (2012).

[7] R. L. Magin, C. Ingo, L. Colon-Perez, W. Triplett and T. H. Mareci, *Microporous Mesoporous Mater* **178**, 39 (2013).

[8] R. Metzler and J. Klafter, *Phys. Rep.* **339**, 1 (2000).

[9] R. Metzler and J. Klafter, *J. Phys. A: Math. Gen.* **37**, R161 (2004).

[10] U. M. B. Marconi, A. Puglisi, L. Rondoni and A. Vulpiani, *Phys. Rep.* **461**, 111 (2008).

[11] U. Balucani, M. H. Lee and V. Tognetti, *Phys. Rep.* **373**, 409 (2003).

[12] I.M. Sokolov, *Soft Matter* **8**, 9043 (2012).

[13] P. C. Bressloff and J. M. Newby, *Rev. Mod. Phys.* **85**, 135 (2013).

[14] R. Metzler, J. H. Jeon, A. G. Cherstvya and E. Barkai,*Phys. Chem. Chem. Phys.* **16**, 24128 (2014).

[15] H. Risken, *The Fokker-Planck Equation*, second ed. (Springer-Verlag, Berlin, 1996).

[16] C. W. Gardiner, *Handbook of Stochastic Methods* (Springer-Verlag, Berlin, 1997).

[17] N. G. Kampen, *Stochastic Processes in Physics and Chemistry* Elsevier, Hungary, 2004.

[18] W. T. Coffey, Y. P. Kalmykov and J. T. Waldron, *The Langevin equation* (World Scientific, Singapore, 2005).

[19] E. W. Montroll and G. H. Weiss, *J. Math. Phys.* **6**, 167 (1965).

[20] R. Kubo, M. Toda, and N. Hashitsume, *Statistical Physics II: Nonequilibrium Statistical Mechanics* (Springer-Verlag, Berlin, 1985).

[21] K. G. Wang and M. Tokuyama, *Physica A* **265**, 341 (1999).

[22]  I. Snook, *The Langevin and Generalised Langevin Approach to the Dynamics of Atomic, Polymeric and Colloidal Systems* (Elsevier, Netherlands, 2007).

[23]  T. D. Frank, *Nonlinear FokkerPlanck Equations* (Springer-Verlag, Berlin, 2005).

[24]  K. S. Fa and K. G. Wang, *Phys. Rev. E* **81**, 011126 (2010).

[25]  B. J. West and S. Picozzi, *Phys. Rev. E* **65**, 037106 (2002).

[26]  K. S. Fa, *Phys. Rev. E* **73**, 061104 (2006).

[27]  S. C. Lim, M. Li and L. P. Teo, *Phys. Lett. A* **372**, 6309 (2008).

[28]  C. H. Eab and S. C. Lim, *Physica A* **389**, 2510 (2010).

# Chapter 2

# Langevin and Fokker-Planck equations

## 2.1 Introduction

The Langevin equation is a very important tool for describing nonequilibrium systems [1–7]. The equation is composed of deterministic terms and a random force. In general, solution for the Langevin equation is diffcult to be obtained, but it can be converted into a differential equation for the probability density function (PDF) which is called the Fokker-Planck equation. Thus, solutions of the Langevin equation are usually obtained from the corresponding Fokker-Planck equation.

## 2.2 General Langevin equation for one variable

The general Langevin equation, for one stochastic variable, has the form

$$\frac{d\xi}{dt} = h_1(\xi, t) + h_2(\xi, t)L(t) , \qquad (2.1)$$

where $\xi$ is a stochastic variable and $L(t)$ is the Langevin force which is assumed to be a Gaussian random variable with the following averages [1,3]:

$$\langle L(t) \rangle = 0 \qquad (2.2)$$

and

$$\langle L(t)L(\bar{t}) \rangle = q\delta(t - \bar{t}), \qquad (2.3)$$

where $\delta(t)$ is the Dirac delta function. The higher correlation functions for $L(t)$ are given by

$$\langle L(t_1)L(t_2)...L(t_{2n-1}) \rangle = 0 \qquad (2.4)$$

and

$$\langle L(t_1)L(t_2)...L(t_{2n})\rangle$$

$$= q^n \left[ \sum_{P_d} \delta(t_{i_1} - t_{i_2})\delta(t_{i_3} - t_{i_4})...\delta(t_{i_{2n-1}} - t_{i_{2n}}) \right], \qquad (2.5)$$

where the sum is over $(2n)!/(2^n n!)$ permutations. The function $h_1(\xi, t)$ is the deterministic drift. Note that the spectral density of the average (2.3), by using the Wiener-Khintchine theorem [1], is independent of the frequency $\omega$, i.e,

$$S(\omega) = 2 \int_{-\infty}^{\infty} e^{-i\omega\tau} \langle L(\tau)L(0)\rangle d\tau = 2q, \qquad (2.6)$$

and it is called white-noise force. When the spectral density is dependent on the frequency $\omega$ the noise is called colored noise (see Appendix 2.7.1). For $h_2(\xi, t)$ which does not depend on $\xi$ the noise term is called additive noise, whereas for $h_2(\xi, t)$ depending on $\xi$ the noise term is called the multiplicative noise. Physically, the additive noise may represent the heat bath acting on the particle of the system, and the multiplicative noise term may represent a fluctuating barrier. For $h_2 = \sqrt{D}$ and $h_1(\xi, t) = 0$, Eq. (2.1) describes the Wiener process and the corresponding probability distribution (probability density function) is described by a Gaussian function. In the case of $h_2(\xi, t)$, some specific functions have been employed to study, for instance, turbulent flows ($h_2(x, t) \sim |x|^a t^b$) [8–10]. It is worth mentioning that when a multiplicative noise term is introduced into the simple Langevin equation (2.1), even separable in time and space, the system can exhibit complex behaviors and a rich variety of processes. In fact, the Langevin equation is a very important tool for describing non-equilibrium systems [1–4]. Moreover, this equation has been extensively investigated; many properties and analytical solutions of it have also been revealed. The Langevin equation is a differential equation for the stochastic variable and its solution, in general, is difficult to be obtained; however it can be transformed into the Fokker-Planck equation which is a differential equation for the distribution function (probability density function), and its solution can be obtained in many cases. It should be noted that the Fokker-Planck equation corresponding to the Langevin equation (2.1) with the white noise involves different orders of prescription in discretization rules for the stochastic integrals [1].

## 2.3 Kramers-Moyal expansion coefficients

The Kramers-Moyal expansion coefficients are defined by

$$D_n(x,t) = \frac{1}{n!} \lim_{\tau \to 0} \frac{1}{\tau} M_n(t)|_{\xi(t)=x}$$

$$= \frac{1}{n!} \lim_{\tau \to 0} \frac{1}{\tau} \left\langle [\xi(t+\tau) - x]^n \right\rangle |_{\xi(t)=x}, \tag{2.7}$$

where $M_n(t)$ is the $n$th moment centered at the initial value $x$ and $\xi(t+\tau)$ for $\tau > 0$ is obtained from Eq. (2.1) with a sharp value $\xi(t) = x$. These coefficients are substituted into the Kramers-Moyal forward expansion (Section 2.4) given by

$$\frac{\partial \rho(x,t)}{\partial t} = \sum_{n=1}^{\infty} \left( -\frac{\partial}{\partial x} \right)^n D_n(x,t)\rho(x,t). \tag{2.8}$$

For $n = 2$ the equation (2.8) is called the Fokker-Planck equation [1]. Now we derive the Kramers-Moyal expansion coefficients related to the Langevin equation (2.1) with the following normalization for the Langevin force:

$$\langle L(t) \rangle = 0 \tag{2.9}$$

and

$$\left\langle L(t)L(\bar{t}) \right\rangle = 2\delta(t - \bar{t}). \tag{2.10}$$

First, we integrate Eq. (2.1), and we obtain

$$\xi(t+\tau) - x = \int_t^{t+\tau} [h_1(\xi(t_1), t_1) + h_2(\xi(t_1), t_1)L(t_1)] \, dt_1. \tag{2.11}$$

Expanding the functions $h_1(\xi(t_1), t_1)$ and $h_2(\xi(t_1), t_1)$ in Taylor series at the point $x$ we arrive at

$$\xi(t+\tau) - x = \int_t^{t+\tau} h_1(x, t_1)dt_1 + \int_t^{t+\tau} (\xi(t_1) - x) \, h_1'(x, t_1)dt_1$$

$$+ \frac{1}{2!} \int_t^{t+\tau} (\xi(t_1) - x)^2 \, h_1'(x, t_1)dt_1 + ... + \int_t^{t+\tau} h_2(x, t_1)L(t_1)dt_1$$

$$+ \int_t^{t+\tau} (\xi(t_1) - x) \, h_2'(x, t_1)L(t_1)dt_1$$

$$+ \frac{1}{2!} \int_t^{t+\tau} (\xi(t_1) - x)^2 \, h_2'(x, t_1)L(t_1)dt_1 + ... , \tag{2.12}$$

where the prime indicates the derivative in relation to $x$. Now we substitute the relation (2.11) inside the integrals of Eq. (2.12), and we obtain

$$\xi(t+\tau) - x = \int_t^{t+\tau} h_1(x, t_1)dt_1 + \int_t^{t+\tau} h_1'(x, t_1) \int_t^{t_1} h_1(x, t_2)dt_2dt_1$$

$$+ \int_t^{t+\tau} h_1'(x, t_1) \int_t^{t_1} h_2(x, t_2)L(t_2)dt_2dt_1 + \int_t^{t+\tau} h_2(x, t_1)L(t_1)dt_1$$

$$+ \int_t^{t+\tau} h_2'(x, t_1)L(t_1) \int_t^{t_1} h_1(x, t_2)dt_2dt_1$$

$$+ \int_t^{t+\tau} h_2'(x, t_1)L(t_1) \int_t^{t_1} h_2(x, t_2)L(t_2)dt_2dt_1 + \dots . \qquad (2.13)$$

The lowest terms are shown in Eq. (2.13). Note that for each Langevin force is accompanied by an integral. The third order of the expansion in Eq. (2.13) contains terms with three Langevin forces which vanish for the average; the next order of the expansion in Eq. (2.13) contains four integrals of the form

$$\int_t^{t+\tau} L(t_1) \int_t^{t_1} L(t_2) \int_t^{t_2} L(t_3) \int_t^{t_3} L(t_4) \dots dt_4dt_3dt_2dt_1, \qquad (2.14)$$

and it can only give a contribution proportional to $\tau^2$ which vanishes for the limit in (2.7). Besides, the two integrals not containing the Langevin force in Eq. (2.13)

$$\int_t^{t+\tau} h_1'(x, t_1) \int_t^{t_1} h_1(x, t_2)dt_2dt_1 = h_1'(x, t)h_1(x, t) \int_t^{t+\tau} \int_t^{t_1} dt_2dt_1 =$$

$$h_1'(x, t)h_1(x, t)\frac{\tau^2}{2}, \qquad (2.15)$$

also give a contribution proportional to $\tau^2$ which vanishes for the limit in (2.7). For any $n$, the integrals not containing the Langevin force in Eq. (2.13) give a contribution proportional to $\tau^n$. Taking the average of Eq. (2.13) and using the correlation functions for the Langevin force (2.2)-(2.5) with $q = 2$, we have

$$D_1(x, t) = \lim_{\tau \to 0} \frac{1}{\tau} \langle [\xi(t+\tau) - x] \rangle |_{\xi(t)=x} = \lim_{\tau \to 0} \frac{1}{\tau} \left[ \int_t^{t+\tau} h_1(x, t_1)dt_1 \right.$$

$$+ \int_t^{t+\tau} h_1'(x, t_1) \int_t^{t_1} h_1(x, t_2)dt_2dt_1 + \int_t^{t+\tau} h_2'(x, t_1) \int_t^{t_1} h_1(x, t_2)dt_2dt_1$$

$$+ 2 \int_t^{t+\tau} h_2'(x, t_1) \int_t^{t_1} h_2(x, t_2) \delta(t_2 - t_1) dt_2 dt_1 \Bigg] . \tag{2.16}$$

Using the following representation for the symmetric $\delta_\epsilon(t)$:

$$\delta_\epsilon(t) = \begin{cases} \frac{1}{\epsilon}, & -\frac{\epsilon}{2} < t < \frac{\epsilon}{2} \\ 0, & \text{otherwise} \end{cases}, \tag{2.17}$$

we obtain

$$D_1(x, t) = h_1(x, t) + h_2(x, t) h_2'(x, t). \tag{2.18}$$

Using the above arguments we obtain the following Kramers-Moyal expansion coefficients for $n > 1$:

$$D_2(x, t) = \frac{1}{2} \lim_{\tau \to 0} \frac{1}{\tau} \left\langle \int_t^{t+\tau} h_2(x, t_1) L(t_1) dt_1 \int_t^{t+\tau} h_2(x, t_2) L(t_2) dt_2 \right\rangle =$$

$$\lim_{\tau \to 0} \frac{h_2^2(x, t)}{\tau} \int_t^{t+\tau} \int_t^{t+\tau} \delta(t_2 - t_1) dt_2 dt_1 =$$

$$\lim_{\tau \to 0} \frac{h_2^2(x, t)}{\tau} \int_t^{t+\tau} dt_1 = h_2^2(x, t) \tag{2.19}$$

and

$$D_n(x, t) = 0, \text{ for } n > 2. \tag{2.20}$$

The coefficient $D_1(x, t)$ is called the drift coefficient, whereas $D_2(x, t)$ is called the diffusion coefficient.

## 2.4 Ito, Stratonovich and other prescriptions

Note that the Langevin equation (2.1), in which the Langevin force is Gaussian distributed and having the $\delta$ correlation function (2.10), is not well defined [1, 3, 4, 11, 12]. To make sense the singular correlation function should have a finite correlation time with a very small width. Besides, the Fokker-Planck equation corresponding to the Langevin equation (2.1) with the white noise involves different orders of prescription in discretization rules for the stochastic integrals [1] is due to the multiplicative noise. To carry out the calculation with the multiplicative noise we use the variable $W(t)$ given by

$$W(t) = \int_0^t L(t_1) dt_1 \tag{2.21}$$

and the following increment:

$$w(\tau) = W(t + \tau) - W(t) = \int_t^{t+\tau} L(t_1)dt_1. \qquad (2.22)$$

The variable $W(t)$ is a much smoother function when compared with $L(t)$, and it even exists if the correlation time of $L(t)$ is zero. In terms of the variable $W(t)$, Eq. (2.1) is known as a Stieltjes integral, i.e.,

$$\xi(t + \tau) - x = \int_t^{t+\tau} h_1(\xi(t_1), t_1)dt_1 + \int_t^{t+\tau} h_2(\xi(t_1), t_1)dW(t_1). \qquad (2.23)$$

Now we derive the Kramers-Moyal expansion coefficients related to the Langevin equation (2.1) for any order of prescription in discretization rules for the stochastic integrals. Note that the distribution of $w(\tau)$ is also Gaussian due to the fact that Eq. (2.22) is linear and $L(t)$ is Gaussian distributed. From Eqs. (2.9) and (2.22) we obtain

$$w(0) = 0 \qquad (2.24)$$

and

$$\langle w(\tau) \rangle = 0. \qquad (2.25)$$

The correlation $\langle w(\tau_2)w(\tau_1) \rangle$ is obtained from Eqs. (2.10) and (2.22)

$$\langle w(\tau_2)w(\tau_1) \rangle = \left\langle \int_t^{t+\tau_2} L(t_2)dt_2 \int_t^{t+\tau_1} L(t_1)dt_1 \right\rangle =$$

$$2\int_t^{t+\tau_2} \int_t^{t+\tau_1} \delta(t_1 - t_2)dt_1 dt_2. \qquad (2.26)$$

Eq. (2.26) is different from zero only for $t_1 = t_2$ due to the $\delta$ function, then the integrations run from $t$ to $t + \tau_1$ or $t + \tau_2$ whatever limit is less; the result is given by

$$\langle w(\tau_2)w(\tau_1) \rangle = \begin{cases} 2\tau_2, & \tau_1 \geq \tau_2 \\ 2\tau_1, & \tau_1 \leq \tau_2 \end{cases}. \qquad (2.27)$$

The higher correlation functions for $w(\tau)$ are obtained from Eqs. (2.5) and (2.22).

The next step is to eliminate the stochastic variable $\xi(t)$ in Eq. (2.23) by using the same procedure given in section 2.3. The iterations of Eq. (2.23) result in the integrals of the form

$$\int_0^\tau \Phi(w(\tau_1), \tau_1)dw(\tau_1), \qquad (2.28)$$

which are not yet defined. The integral (2.28) may be written as a sum and its result depends on the choice of the point of $\Phi(w(\tau_i), \tau_i)$. For instance, the Ito prescription is chosen at the last point $\tau_i$ of the increment $w(\tau_{i+1}) - w(\tau_i)$, whereas the Stratonovich prescription is chosen at the middle point $(w(\tau_{i+1}) + w(\tau_i))/2$. We may specify any point within the increment $\Delta w(\tau_i) = w(\tau_{i+1}) - w(\tau_i)$ as follows:

$$w(\tau_i) + \lambda \Delta w(\tau_i) = (1 - \lambda)w(\tau_i) + \lambda w(\tau_{i+1}), \qquad (2.29)$$

where $0 \leq \lambda \leq 1$. For $\lambda = 0$ we have the Ito prescription, whereas the Stratonovich prescription the value of $\lambda$ is $\lambda = 1/2$. Hence the integral (2.28) leads to

$$\int_0^\tau \Phi(w(\tau_i), \tau_1)dw(\tau_1) =$$

$$\lim_{\Delta \to 0} \sum_{i=0}^{N-1} \Phi(\lambda w(\tau_{i+1}) + (1 - \lambda)w(\tau_i), \tau_i)\left[w(\tau_{i+1}) - w(\tau_i)\right], \qquad (2.30)$$

where

$$\Delta = \max(\tau_{i+1} - \tau_i); \quad 0 = \tau_0 < \tau_1, ... < \tau_N. \qquad (2.31)$$

Following the expansions given in the previous section and using the relations (2.25) and (2.27) we obtain the following Kramers-Moyal expansion coefficients:

$$D_1(x, t) = \lim_{\tau \to 0} \frac{1}{\tau} \times$$

$$\left[\tau h_1(x, t + \theta_1\tau) + h_2(x, t + \theta_3\theta_2\tau)h_2'(x, t + \theta_3\tau)\left\langle \int_0^\tau w(\tau_1)dw(\tau_1)\right\rangle\right]$$

$$= h_1(x, t) + \lim_{\tau \to 0} \frac{1}{\tau}h_2(x, t + \theta_3\theta_2\tau)h_2'(x, t + \theta_3\tau)$$

$$\times \left\langle \sum_{i=0}^{N-1} \left[\lambda w(\tau_{i+1}) + (1 - \lambda)w(\tau_i)\right]\left[w(\tau_{i+1}) - w(\tau_i))\right]\right\rangle = h_1(x, t)$$

$$+ \lim_{\tau \to 0} \frac{1}{\tau}h_2(x, t+\theta_3\theta_2\tau)h_2'(x, t+\theta_3\tau) \sum_{i=0}^{N-1} \left[2\lambda(\tau_{i+1} - \tau_i) + 2(1 - \lambda)(\tau_i - \tau_i))\right]$$

$$= h_1(x, t) + 2\lambda \lim_{\tau \to 0} \frac{1}{\tau}h_2(x, t + \theta_3\theta_2\tau)h_2'(x, t + \theta_3\tau) \sum_{i=0}^{N-1} (\tau_{i+1} - \tau_i)$$

$$= h_1(x,t) + 2\lambda \lim_{\tau \to 0} \frac{1}{\tau} h_2(x, t + \theta_3\theta_2\tau) h_2'(x, t + \theta_3\tau)\tau$$

$$= h_1(x,t) + 2\lambda h_2(x,t) h_2'(x,t) \tag{2.32}$$

and

$$D_2(x,t) = \frac{1}{2} \lim_{\tau \to 0} \frac{1}{\tau} h_2^2(x, t + \theta_2\tau) \langle w^2(\tau) \rangle = h_2^2(x,t), \tag{2.33}$$

where $0 \leq \theta_i \leq 1$. It should be noted that the Stratonovich prescription is consistent with Eqs. (2.18) and (2.19). Moreover, the coefficient $D_1(x,t)$ given by Eq. (2.32) contains a spurious drift term $2\lambda h_2(x,t)h_2'(x,t)$, and the spurious drift is missing only for the Ito prescription ($\lambda = 0$). In the case of additive noise $\Phi$ does not depend on the choice of the point $\tau_i$ of $w(t)$, then the integral (2.28) gives the same result for all of the prescriptions. Besides, solutions for different prescriptions could describe different behaviors, then further information of the microscopic structure of a system is necessary in order to choose which of the prescriptions is the correct one or it must be determined on the basis of the available experimental data [13, 14].

## 2.5    Kramers-Moyal expansion and Fokker-Planck equation

To derive a differential equation for the PDF which connects with the Kramers-Moyal coefficients, obtained in Section (2.4), we first consider that the probability density $\rho(x, t + \tau)$ at time $t + \tau$ is obtained from the probability density $\rho(x, t + \tau; x', t)$ by

$$\rho(x, t + \tau) = \int \rho(x, t + \tau; x', t)dx', \tag{2.34}$$

where $\tau \geq 0$ and it is small. Eq. (2.34) can be written in terms of the transition probability $P(x, t + \tau | x', t)$ [1] as

$$\rho(x, t + \tau) = \int P(x, t + \tau | x', t)\rho(x', t)dx'. \tag{2.35}$$

Besides, the transition probability and the moments $M_n$ can be connected with the characteristic function as follows:

$$C(u, x', t, \tau) = \left\langle e^{iu(x-x')} \right\rangle = \int_{-\infty}^{\infty} e^{iu(x-x')} P(x, t + \tau | x', t)dx =$$

$$\int_{-\infty}^{\infty} \sum_{n=o}^{\infty} \frac{[iu(x - x')]^n}{n!} P(x, t+\tau | x', t)dx = 1 + \sum_{n=1}^{\infty} \frac{[iu]^n M_n(x', t, \tau)}{n!}. \tag{2.36}$$

Note that the characteristic function is the Fourier transform of the transition probability, then the transition probability is the inverse Fourier transform of the characteristic function

$$P(x, t + \tau | x', t) = \frac{1}{2\pi} \int_{-\infty}^{\infty} e^{-iu(x-x')} C(u, x', t, \tau) du. \qquad (2.37)$$

Substituting the result (2.36) into (2.37) we have

$$P(x, t + \tau | x', t) = \frac{1}{2\pi} \sum_{n=0}^{\infty} \left( -\frac{\partial}{\partial x} \right)^n \int_{-\infty}^{\infty} e^{-iu(x-x')} M_n(x', t, \tau) du. \qquad (2.38)$$

Using

$$\delta(z) = \frac{1}{2\pi} \int_{-\infty}^{\infty} e^{-iuz} du \quad \text{and} \quad \delta(x - x') f(x') = \delta(x - x') f(x) \qquad (2.39)$$

we obtain from Eq. (2.38) the following result:

$$P(x, t + \tau | x', t) = \sum_{n=0}^{\infty} \left( -\frac{\partial}{\partial x} \right)^n M_n(x, t, \tau) \delta(x - x'). \qquad (2.40)$$

Now substituting Eq. (2.40) into Eq. (2.35) yields

$$\rho(x, t + \tau) = \int \sum_{n=0}^{\infty} \left( -\frac{\partial}{\partial x} \right)^n M_n(x, t, \tau) \delta(x - x') \rho(x', t) dx'$$

$$= \rho(x, t) + \sum_{n=1}^{\infty} \left( -\frac{\partial}{\partial x} \right)^n \frac{M_n(x, t, \tau)}{n!} \rho(x, t) dx'. \qquad (2.41)$$

Expanding $M_n(x, t, \tau)$ in Taylor series we obtain

$$\rho(x, t + \tau) - \rho(x, t)$$

$$= \sum_{n=1}^{\infty} \left( -\frac{\partial}{\partial x} \right)^n \left[ \frac{M_n(x, t)}{n!} + \tau D_n(x, t) + O(\tau^2) \right] \rho(x, t). \qquad (2.42)$$

The initial value for the transition probability is assumed to be

$$P(x, t | x', t) = \delta(x - x'). \qquad (2.43)$$

With the initial value (2.43) we obtain

$$M_n(x', t) = \int (x - x')^n P(x, t | x', t) dx = \int (x - x')^n \delta(x - x') dx = 0. \qquad (2.44)$$

By taking into account the result (2.44) and the limiting process $\tau \to 0$ (only for the linear terms in $\tau$) we obtain from Eq. (2.42) the following differential equation:

$$\frac{\partial \rho(x, t)}{\partial t} = \sum_{n=1}^{\infty} \left( -\frac{\partial}{\partial x} \right)^n D_n(x, t) \rho(x, t) = L_{KM}(x, t) \rho(x, t), \qquad (2.45)$$

where

$$L_{KM}(x,t) = \sum_{n=1}^{\infty} \left(-\frac{\partial}{\partial x}\right)^n D_n(x,t) \qquad (2.46)$$

is called the Kramers-Moyal operator.

Substituting Eq. (2.35) into Eq. (2.45) with $t$ replaced by $t'$ and $t + \tau$ replaced by $t$ yields

$$\frac{\partial P(x,t|x',t')}{\partial t} = L_{KM}(x,t)P(x,t|x',t'). \qquad (2.47)$$

Thus, the evolution of the transition probability also obeys the differential equation (2.45).

## 2.6   General Langevin equation for several variables

The general Langevin equation for $N$ stochastic variables $\{\xi\} = \{\xi_1, \xi_2, ..., \xi_N\}$ [1] is described by

$$\frac{d\xi_i}{dt} = h_i(\{\xi\},t) + h_{ij}(\{\xi\},t)L_j(t) , \quad i = 1, 2, ..., N, \qquad (2.48)$$

with

$$\langle L_j(t)\rangle = 0 \qquad (2.49)$$

and

$$\langle L_i(t)L_j(\bar{t})\rangle = 2\delta_{ij}\delta(t - \bar{t}), \qquad (2.50)$$

where $\delta_{ij}$ is the Kronecker delta defined by

$$\delta_{ij} = \begin{cases} 1 \text{ if } & i = j \\ 0 \text{ if } & i \neq j \end{cases} . \qquad (2.51)$$

In Eq. (2.48) the summation is performed over indice $j$ appearing twice.

The Kramers-Moyal expansion coefficients related to the nonlinear Langevin equation for one and several stochastic variables are obtained similarly. Following the procedure given in the previous sections yields

$$D_i(\{\mathbf{x}\},t) = \lim_{\tau \to 0} \frac{1}{\tau} \langle [\xi_i(t+\tau) - x_i]\rangle |_{\xi_k(t)=x_k}$$

$$= h_i(\{\mathbf{x}\},t) + 2\lambda h_{lj}(\{\mathbf{x}\},t)\frac{\partial}{\partial x_l}h_{ij}(\{\mathbf{x}\},t), \quad k = 1, 2, ....., N, \qquad (2.52)$$

$$D_{ij}(\{\mathbf{x}\},t) = \frac{1}{2}\lim_{\tau \to 0}\frac{1}{\tau}\langle [\xi_i(t+\tau) - x_i][\xi_j(t+\tau) - x_j]\rangle |_{\xi_k(t)=x_k}$$

$$= h_{il}(\{\mathbf{x}\}, t) h_{jl}(\{\mathbf{x}\}, t), \quad k = 1, 2, \ldots, N \tag{2.53}$$

and

$$D_{i_1 i_2 \ldots i_n}(\{\mathbf{x}\}, t) = \frac{1}{n!} \lim_{\tau \to 0} \frac{1}{\tau} \langle [\xi_{i_1}(t + \tau) - x_{i_1}] \ldots [\xi_{i_n}(t + \tau) - x_{i_n}] \rangle |_{\xi_k(t) = x_k}$$

$$= 0, \quad \text{for } n \geq 3. \tag{2.54}$$

The Fokker-Planck equation for several variables is given by

$$\frac{\partial \rho(\{\mathbf{x}\}, t)}{\partial t} = \left[ -\frac{\partial}{\partial x_i} D_i(\{\mathbf{x}\}, t) + \frac{\partial^2}{\partial x_i x_j} D_{ij}(\{\mathbf{x}\}, t) \right] \rho(\{\mathbf{x}\}, t). \tag{2.55}$$

Note that, in the case of several variables, the Fokker-Planck equation (2.55) may not depend on any stochastic prescription for the multiplicative white noise. As an example we consider the following Langevin equation [15]:

$$\frac{d^2 x}{dt^2} = -\gamma_0 \frac{dx}{dt} - V'(x) + g(x)L(t), \tag{2.56}$$

where $V(x)$ is an external potential. In order to apply the above scheme we write Eq. (2.56) in terms of the first derivatives, i.e, we transform the second-order differential equation (2.56) into two differential equations of the first order as follows:

$$\frac{dx}{dt} = v, \tag{2.57}$$

$$\frac{dv}{dt} = -\gamma_0 v - V'(x) + g(x)L(t). \tag{2.58}$$

Identifying $\xi_1 = x$ and $\xi_2 = v$ we have from Eqs. (2.52) and (2.53) the following results for the coefficients $D_i$ and $D_{ij}$:

$$D_1 = v, \quad D_2 = -\gamma_0 v - V'(x), D_{11} = D_{12} = D_{21} = 0 \tag{2.59}$$

and

$$D_{22} = g^2(x). \tag{2.60}$$

Substituting the coefficients (2.59) and (2.60) into Eq. (2.55) yields

$$\frac{\partial \rho(x, v, t)}{\partial t} = \left[ -v \frac{\partial}{\partial x} + \frac{\partial}{\partial v} (\gamma_0 v + V'(x)) + g^2(x) \frac{\partial^2}{\partial v^2} \right] \rho(x, v, t). \tag{2.61}$$

One can see that the prescription parameter $\lambda$ does not appear in Eq. (2.61), and Eq. (2.61) is the Klein-Kramers equation with a non-constant diffusion coefficient. This means that the inertial term may avoid the prescription problem induced by the multiplicative white noise in the Langevin equation without the presence of the inertial term.

## 2.7 Appendices

### 2.7.1 *Colored noise*

A usual way to deal with a colored noise is to obtain its corresponding stochastic force by means of a Langevin equation. In this approach the procedure to deal with a colored noise is to extend the space of variables so that the noise itself becomes a variable driven by the white noise. For instance, the Gaussian stochastic force with zero mean and an exponential correlation function given by

$$\langle \eta_1(t) \rangle = 0 \tag{2.62}$$

and

$$\langle \eta_1(t)\eta_1(\bar{t}) \rangle = D\gamma e^{-\gamma|t-\bar{t}|}, \tag{2.63}$$

can be obtained from the following Langevin equation:

$$\frac{d\eta_1}{dt} = -\gamma\eta_1(t) + \gamma L(t), \tag{2.64}$$

where the Langevin force $L(t)$ is described by Eqs. (2.2) and (2.3). Eq. (2.64) describes the Ornstein-Uhlenbeck process. The exponential correlation (2.63) has been applied to diverse systems, including nonlinear dynamics [5,16]. The noise $\eta_1(t)$ reduces to the white noise in the limit $\gamma \to \infty$. We now solve Eq. (2.64) and to prove the equivalence of the above equations. Integrating Eq. (2.64) yields

$$\eta_1(t) = \gamma e^{-\gamma t} \int_0^t e^{\gamma t_1} L(t_1) dt_1, \tag{2.65}$$

with $\eta_1(0) = 0$. Averaging Eq. (2.65) and then using Eq. (2.2) results in Eq. (2.62). From Eq. (2.65) we obtain the following correlation function:

$$\langle \eta_1(t)\eta_1(\bar{t}) \rangle = \gamma^2 e^{-\gamma(t+\bar{t})} \left\langle \int_0^t e^{\gamma t_1} L(t_1) dt_1 \int_0^{\bar{t}} e^{\gamma t_2} L(t_2) dt_2 \right\rangle$$

$$= q\gamma^2 e^{-\gamma(t+\bar{t})} \int_0^t \int_0^{\bar{t}} e^{\gamma(t_1+t_2)} \delta(t_1 - t_2) dt_1 dt_2 \tag{2.66}$$

Eq. (2.66) is different from zero only for $t_1 = t_2$ due to the $\delta$ function, then the integrations run from 0 to $t$ or $\bar{t}$ whatever limit is less. Thus we have

$$\langle \eta_1(t)\eta_1(\bar{t}) \rangle$$

$$= \begin{cases} q\gamma^2 e^{-\gamma(t+\bar{t})} \int_0^{\bar{t}} e^{2\gamma t_1} dt_1 = \frac{q\gamma}{2} \left[ e^{-\gamma(t-\bar{t})} - e^{-\gamma(t+\bar{t})} \right], t > \bar{t} \\ q\gamma^2 e^{-\gamma(t+\bar{t})} \int_0^{t} e^{2\gamma t_1} dt_1 = \frac{q\gamma}{2} \left[ e^{-\gamma(\bar{t}-t)} - e^{-\gamma(t+\bar{t})} \right], t < \bar{t} \end{cases} . \quad (2.67)$$

Eq. (2.67) can be written as follows:

$$\langle \eta_1(t)\eta_1(\bar{t}) \rangle = \frac{q\gamma}{2} \left[ e^{-\gamma|t-\bar{t}|} - e^{-\gamma(t+\bar{t})} \right]. \quad (2.68)$$

Eq. (2.68) leads to Eq. (2.63) in the stationary state, i.e, for $q = 2D$ and $t \gg 1$ we obtain

$$\langle \eta_1(t)\eta_1(\bar{t}) \rangle = D\gamma e^{-\gamma|t-\bar{t}|}. \quad (2.69)$$

The parameter

$$\tau = \frac{\int_0^\infty \bar{\tau} \langle \eta_1(t)\eta_1(t+\bar{\tau}) \rangle d\bar{\tau}}{\int_0^\infty \langle \eta_1(t)\eta_1(t+\bar{\tau}) \rangle d\bar{\tau}} \quad (2.70)$$

gives

$$\tau_{\text{OUN}} = \frac{1}{\gamma}. \quad (2.71)$$

which has the meaning of the relaxation time of the correlation.

Another popular Gaussian stochastic force is the harmonic noise [17–19] and it is described by the following Langevin equation:

$$\frac{d^2\eta_2}{dt^2} + \gamma\frac{d\eta_2}{dt} + \Omega_0^2\eta_2 = \sqrt{2\epsilon}\Omega_0^2 L(t), \quad (2.72)$$

where the Langevin force $L(t)$ is described by Eqs. (2.2) and (2.3). Note that the difference between Eq. (2.64) and Eq. (2.72) is the inertial term, and it can produce oscillating solution. The solution can be obtained from the Laplace transform; applying the Laplace transform to Eq. (2.72) yields

$$\eta_{s2}(s) = G_s(s) \left[ v(0) + \gamma\eta_2(0) + \eta_2(0)s + \sqrt{2\epsilon}\Omega_0^2 L_s(s) \right], \quad (2.73)$$

where

$$G_s(s) = \frac{1}{\Omega_0^2 + \gamma s + s^2}. \quad (2.74)$$

Applying the inverse Laplace transform to Eq. (2.73) we arrive at

$$\eta_2(t) = [v(0) + \gamma\eta_2(0)] G(t)$$

$$+ \eta_2(0)\frac{dG(t)}{dt} + \sqrt{2\epsilon}\Omega_0^2 \int_0^t L(\tau)G(t-\tau)d\tau, \quad (2.75)$$

with $G(0) = 0$. The solution for $G(t)$ is given by

$$G(t) = \begin{cases} \frac{e^{-\frac{\gamma t}{2}}\sinh(\omega_2 t)}{\omega_2}, & \Omega_0^2 < \frac{\gamma^2}{4} \\ te^{-\frac{\gamma t}{2}}, & \Omega_0^2 = \frac{\gamma^2}{4} \\ \frac{e^{-\frac{\gamma t}{2}}\sin(\omega_1 t)}{\omega_1}, & \Omega_0^2 > \frac{\gamma^2}{4} \end{cases} \tag{2.76}$$

where

$$\omega_1^2 = \Omega_0^2 - \frac{\gamma^2}{4} \tag{2.77}$$

and

$$\omega_2^2 = \frac{\gamma^2}{4} - \Omega_0^2. \tag{2.78}$$

Averaging Eq. (2.75) and then using Eq. (2.2) yields

$$\langle \eta_2(t) \rangle = 0, \tag{2.79}$$

in the long–time limit.

For the correlation function we have

$$\langle \eta_2(t)\eta_2(t+\tau) \rangle =$$

$$2\epsilon\Omega_0^4 \left\langle \int_0^t L(t_1)G(t-t_1)dt_1 \int_0^{t+\tau} L(t_2)G(t+\tau-t_2)dt_2 \right\rangle$$

$$= 2\epsilon\Omega_0^4 \int_0^t \int_0^{t+\tau} G(t-t_1)G(t+\tau-t_2)\delta(t_1-t_2)dt_2 dt_1$$

$$= 2\epsilon\Omega_0^4 \int_0^t G(t-t_1)G(t+\tau-t_1)dt_1, \tag{2.80}$$

with $q = 1$.

In the case of $\Omega_0^2 > \frac{\gamma^2}{4}$ we have

$$\langle \eta_2(t)\eta_2(t+\tau) \rangle$$

$$= \frac{2\epsilon\Omega_0^4}{\omega_1^2} \int_0^t e^{-\gamma\left(t+\frac{\tau}{2}-t_1\right)} \sin\left(\omega_1(t_1-t)\right) \sin\left(\omega_1(t_1-t-\tau)\right) dt_1$$

$$= \frac{\epsilon\Omega_0^2 e^{-\frac{\gamma\tau}{2}}}{\gamma} \left\{ \cos\left(\omega_1\tau\right) + \frac{\gamma}{2\omega_1}\sin\left(\omega_1\tau\right) + \frac{e^{-\gamma t}}{\omega_1^2} \right.$$

$$\times \left. \left[ \frac{\gamma^2}{4}\cos\left(\omega_1\left(2t+\tau\right)\right) - \Omega_0^2\cos\left(\omega_1\tau\right) - \frac{\gamma\omega_1}{2}\sin\left(\omega_1\left(2t+\tau\right)\right) \right] \right\}. \tag{2.81}$$

For $t \gg 1$, Eq. (2.81) reduces to

$$\langle \eta_2(t)\eta_2(t+\tau) \rangle = \frac{\epsilon\Omega_0^2 e^{-\frac{\gamma\tau}{2}}}{\gamma} \left[ \cos(\omega_1\tau) + \frac{\gamma}{2\omega_1} \sin(\omega_1\tau) \right]. \qquad (2.82)$$

The result (2.82) is the stationary state because it depends only on the time difference $\tau$.

The parameter (2.70) for the harmonic noise is given by

$$\tau_{\text{HN}} = \frac{\gamma}{\Omega_0^2} - \frac{1}{\gamma}, \quad \Omega_0^2 > \frac{\gamma^2}{4}. \qquad (2.83)$$

We see that $\tau_{\text{HN}}$ may assume positive and negative values. In this case the parameter $\tau_{\text{HN}}$ looses the meaning of the relaxation time of the correlation.

For $\Omega_0^2 = \frac{\gamma^2}{4}$ the correlation function can be obtained from the solution (2.81) by taking the limit $\omega_1 \to 0$, and the result is given by

$$\langle \eta_2(t)\eta_2(t+\tau) \rangle$$

$$= \epsilon\Omega_0^2 \left[ \frac{1}{\gamma} + \frac{\tau}{2} - \left( \frac{1}{\gamma} + \frac{\tau}{2} + t + \frac{\gamma t(t+\tau)}{2} \right) e^{-\gamma t} \right] e^{-\frac{\gamma\tau}{2}}. \qquad (2.84)$$

The stationary solution is obtained by taking $t \to \infty$, i.e.,

$$\langle \eta_2(t)\eta_2(t+\tau) \rangle = \epsilon\Omega_0^2 \left( \frac{1}{\gamma} + \frac{\tau}{2} \right) e^{-\frac{\gamma\tau}{2}}. \qquad (2.85)$$

The parameter (2.70) for the harmonic noise is given by

$$\tau_{\text{HN}} = \frac{3}{\gamma}, \quad \Omega_0^2 = \frac{\gamma^2}{4}. \qquad (2.86)$$

One can see that $\tau_{\text{HN}}$ is three times the $\tau_{\text{OUN}}$.

For $\Omega_0^2 < \frac{\gamma^2}{4}$ we can also obtain the correlation function from the solution (2.81). In this case we take

$$\omega_1 = \sqrt{\Omega_0^2 - \frac{\gamma^2}{4}} = i\omega_2. \qquad (2.87)$$

Substituting Eq. (2.87) into Eq. (2.81) yields

$$\langle \eta_2(t)\eta_2(t+\tau) \rangle = \frac{\epsilon\Omega_0^2 e^{-\frac{\gamma\tau}{2}}}{\gamma}$$

$$\times \left\{ \cosh(\omega_2\tau) + \frac{\gamma}{2\omega_2} \sinh(\omega_2\tau) - \frac{e^{-\gamma t}}{\omega_2^2} \left[ \frac{\gamma^2}{4} \cosh(\omega_2(2t+\tau)) \right. \right.$$

$$\left. \left. -\Omega_0^2 \cosh(\omega_2\tau) + \frac{\gamma\omega_2}{2} \sinh(\omega_2(2t+\tau)) \right] \right\}. \qquad (2.88)$$

The stationary solution $(t \to \infty)$ is given by

$$\langle \eta_2(t)\eta_2(t+\tau) \rangle = \frac{\epsilon \Omega_0^2}{\gamma} \left[ \cosh{(\omega_2 \tau)} + \frac{\gamma}{2\omega_2} \sinh{(\omega_2 \tau)} \right] e^{-\frac{\gamma \tau}{2}}. \qquad (2.89)$$

The parameter (2.70) for the harmonic noise is given by

$$\tau_{\mathrm{HN}} = \frac{\gamma}{\Omega_0^2} - \frac{1}{\gamma}, \quad \Omega_0^2 < \frac{\gamma^2}{4}. \qquad (2.90)$$

Note that the $\tau_{\mathrm{HN}}$ has the same result for both cases $\Omega_0^2 > \frac{\gamma^2}{4}$ and $\Omega_0^2 < \frac{\gamma^2}{4}$. However, only the first case $\Omega_0^2 > \frac{\gamma^2}{4}$ may assume positive and negative values due to the oscillatory behavior of the noise.

## Bibliography

[1]  H. Risken, *The Fokker-Planck Equation*, second ed. (Springer-Verlag, Berlin, 1996).

[2]  C. W. Gardiner, *Handbook of Stochastic Methods* (Springer-Verlag, Berlin, 1997).

[3]  N. G. Kampen, *Stochastic Processes in Physics and Chemistry* Elsevier, Hungary, 2004.

[4]  W. T. Coffey, Y. P. Kalmykov and J. T. Waldron, *The Langevin equation* (World Scientific, Singapore, 2005).

[5]  M. Gitterman, *The Noisy Oscillator* (World Scientific, Singapore, 2005).

[6]  I. Snook, *The Langevin and Generalised Langevin Approach to the Dynamics of Atomic, Polymeric and Colloidal Systems* (Elsevier, Netherlands, 2007).

[7]  M. D. Haw, *J. Phys.: Cond. Matt.* **14**, 7769 (2002).

[8]  L. F. Richardson, *Proc. R. Soc. London*, Ser. A **110**, 709 (1926).

[9]  G. K. Batchelor, *Proc. Cambridge Philos. Soc.* **48**, 345 (1952).

[10]  H. G. E. Hentschel and I. Procaccia, *Phys. Rev. A* **29**, 1461 (1984).

[11]  Y. L. Klimontovich, *Physica A* **163**, 515 (1990).

[12]  R. Mannella and P. V. E. McClintock, *Fluctuation and Noise Lett.* **11**, 1240010 (2012).

[13]  G. Pesce , A. McDaniel , S. Hottovy , J. Wehr and G. Volpe *Nature Commun.* **4**, 2733 (2013).

[14]  G. Volpe and J. Wehr, *Rep. Prog. Phys.* **79**, 053901 (2016).

[15]  J. M. Sancho, *Phys. Rev. E* **92**, 062110 (2015).

[16]  F. Moss, P. V. E. McClintock (eds.), *Noise in nonlinear dynamical systems*, Vol. 1-3 (Cambridge University Press, Cambridge, 1989).

[17]  L. Schimansky-Geier and Ch. Ziilicke, *Z. Phys. B* **79**, 451 (1990).

[18]  J. J. Hesse and L. Schimansky-Geier, *Z. Phys. B* **84**, 467 (1991).

[19]  A. Neiman and L. Schimansky-Geier, *Phys. Rev. Lett.* **72**, 2988 (1994).

# Chapter 3

# Fokker-Planck equation for one variable and its solution

## 3.1 Introduction

In this chapter, we consider some methods for solving the one-variable Fokker-Planck equation obtained from the Langevin equation with multiplicative white noise. A few applications are also discussed. The Fokker-Planck equation for the probability distribution corresponding to the Langevin equation (2.1) is given by

$$\frac{\partial \rho(x,t)}{\partial t} = -\frac{\partial}{\partial x}\left[D_1(x,t)\rho(x,t)\right] + \frac{\partial^2}{\partial x^2}\left[D_2(x,t)\rho(x,t)\right] , \qquad (3.1)$$

where $D_1(x,t)$ and $D_2(x,t)$ are the drift and diffusion coefficients given by

$$D_1(x,t) = h_1(x,t) + 2\lambda \frac{\partial h_2(x,t)}{\partial x} h_2(x,t) \qquad (3.2)$$

and

$$D_2(x,t) = h_2^2(x,t), \qquad (3.3)$$

and $\lambda$ ($0 \leq \lambda \leq 1$) is the prescription parameter due to the discretization rules for the stochastic integrals. In particular, $\lambda = 0$ corresponds to the Ito prescription, $\lambda = 1/2$ corresponds to the Stratonovich prescription and $\lambda = 1$ corresponds to the transport prescription (it is also called kinetic or Hänggi-Klimontovich prescription) [1–4]. Notice that Eq. (3.2) does not have a spurious drift only for the Ito prescription, and the prescription parameter is directly linked to the coefficient $h_2(x,t)$ of the fluctuation. Besides, Eq. (3.1) corresponding to the Ito, Stratonovich and transport prescriptions can be written as follows:

$$\frac{\partial \rho(x,t)}{\partial t} = -\frac{\partial}{\partial x}\left[h_1(x,t)\rho(x,t)\right] + \frac{\partial^2}{\partial x^2}\left[h_2^2(x,t)\rho(x,t)\right] , \quad \lambda = 0, \qquad (3.4)$$

$$\frac{\partial \rho(x,t)}{\partial t} = -\frac{\partial}{\partial x}\left[h_1(x,t)\rho(x,t)\right]$$

$$+ \frac{\partial}{\partial x}\left[h_2(x,t)\frac{\partial h_2(x,t)\rho(x,t)}{\partial x}\right], \quad \lambda = \frac{1}{2} \tag{3.5}$$

and

$$\frac{\partial \rho(x,t)}{\partial t} = -\frac{\partial}{\partial x}\left[h_1(x,t)\rho(x,t)\right] + \frac{\partial}{\partial x}\left[h_2^2(x,t)\frac{\partial \rho(x,t)}{\partial x}\right], \quad \lambda = 1. \tag{3.6}$$

In the Ito prescription the multiplicative noise term $h_2^2(x,t)$ is inside the second derivative, whereas in the Stratonovich and transport prescriptions the term $h_2^2(x,t)$ and the derivative are placed in different positions. For any prescription and $h_2(x,t) > 0$, Eq. (3.1) can be written as follows:

$$\frac{\partial \rho(x,t)}{\partial t} = -\frac{\partial}{\partial x}\left[h_1(x,t)\rho(x,t)\right]$$

$$+ \frac{\partial}{\partial x}\left[h_2^{2\lambda}(x,t)\frac{\partial h_2^{2(1-\lambda)}(x,t)\rho(x,t)}{\partial x}\right]. \tag{3.7}$$

## 3.2 Time-independent drift and diffusion coefficients

In this section, we consider time-independent drift and diffusion coefficients for obtaining stationary and nonstationary solutions of the Fokker-Planck equation (3.1). Before obtaining expressions for the solutions, we note that the Fokker-Planck equation (3.1) may be written as a continuity equation:

$$\frac{\partial \rho(x,t)}{\partial t} + \frac{\partial S(x,t)}{\partial x} = 0 , \tag{3.8}$$

where $S(x,t)$ is the probability current given by

$$S(x,t) = \left[D_1(x,t) - \frac{\partial}{\partial x}D_2(x,t)\right]\rho(x,t) . \tag{3.9}$$

Eq. (3.8) may be interpreted as follows. Integrating Eq. (3.8) with respect to $x$ yields

$$S(x,t)|_{x=x_1} - S(x,t)|_{x=x_2} = \frac{\partial}{\partial t}\int_{x_1}^{x_2}\rho(x,t)dx , \tag{3.10}$$

Eq. (3.10) is illustrated schematically in Fig. 3.1. On the one hand the probability current gives the probability crossing a given point per unit time, then the left-hand side of Eq. (3.10) is equal to the probability flowing into the region $[x_1, x_2]$ per unit time at $x_1$, minus the probability flowing

Fig. 3.1   Illustration of the one-dimensional continuity equation.

out per unit time at $x_2$. On the other hand the integral of the right-hand side of Eq. (3.10) is the probability contained in the region $[x_1, x_2]$, so the right-hand side gives the change in probability in the region per unit time. Therefore, Eq. (3.10) express the conservation of the probability.

For time-independent drift and diffusion coefficients, and with the following transformations:

$$\bar{D}_2(x) = D = \left(\frac{dy}{dx}\right)^2 D_2(x), \tag{3.11}$$

$$\bar{D}_1(x) = \frac{dy}{dx}D_1(x) + \frac{d^2y}{dx^2}D_2(x) = \sqrt{\frac{D}{D_2(x)}}\left[D_1(x) - \frac{1}{2}\frac{dD_2(x)}{dx}\right] \tag{3.12}$$

and

$$\bar{\rho} = J\rho = \left(\frac{dy}{dx}\right)^{-1} = \sqrt{\frac{D_2(x)}{D}}\rho, \tag{3.13}$$

the Fokker-Planck equation

$$\frac{\partial \rho(x,t)}{\partial t} = \left[-\frac{\partial}{\partial x}D_1(x) + \frac{\partial^2}{\partial x^2}D_2(x)\right]\rho(x,t) , \tag{3.14}$$

can be transformed into an equation with constant diffusion coefficient $D > 0$, i.e,

$$\frac{\partial \bar{\rho}(y,t)}{\partial t} = \left[-\frac{\partial}{\partial y}\bar{D}_1(y) + D\frac{\partial^2}{\partial y^2}\right]\bar{\rho}(y,t), \tag{3.15}$$

where $J$ is the Jacobian of transformation and

$$y = y(x) = \int_{x_0}^{x}\sqrt{\frac{D}{D_2(u)}}du. \tag{3.16}$$

### 3.2.1  *Stationary solution*

We now consider time-independent drift and diffusion coefficients for obtaining stationary solutions of the Fokker-Planck equation (3.8). By setting $S(x,t) = S$, where S is a constant, we have

$$S = D_1(x)\rho_{st}(x) - \frac{d}{dx}[D_2(x)\rho_{st}(x)] \ . \qquad (3.17)$$

Solution of Eq. (3.17) is obtained as follows. Multiplying Eq. (3.17) by $\exp(-f(x))$ in both sides and after some manipulations we obtain

$$\frac{d}{dx}\left[e^{-f(x)}D_2(x)\rho_{st}(x)\right]$$

$$= e^{-f(x)}\left[-\frac{df(x)}{dx}D_2(x)\rho_{st}(x) + D_1(x)\rho_{st}(x) - S\right] \ . \qquad (3.18)$$

Put $df(x)/dx = D_1(x)/D_2(x)$, and we arrive at

$$f(x) = \int^x \frac{D_1(x')}{D_2(x')}dx' \qquad (3.19)$$

and

$$\frac{d}{dx}\left[e^{-f(x)}D_2(x)\rho_{st}(x)\right] = -Se^{-f(x)} \ . \qquad (3.20)$$

Integrating Eq. (3.20) yields

$$\rho_{st}(x) = e^{-\Phi(x)}\left[C_0 - S\int^x \frac{\exp(\Phi(x'))}{D_2(x')}dx'\right], \qquad (3.21)$$

where $C_0$ is an integration constant and $\Phi(x)$ is the potential given by

$$\Phi(x) = \ln\left(D_2(x)\right) - \int^x \frac{D_1(x')}{D_2(x')}dx'. \qquad (3.22)$$

The solution (3.21) has two constants: One of them is determined by the normalization, whereas the other one is determined from the boundary conditions.

In particular, for natural boundary conditions (where $x$ extends to $\pm\infty$) the PDF and the probability current must vanish at $\pm\infty$, and therefore $S$ must be zero for any value of $x$. For this case, the stationary solution reduces to

$$\rho_{st}(x) = C_0 \exp\left(-\Phi(x)\right), \qquad (3.23)$$

and $C_0$ is determined by the normalization.

### 3.2.2 Method of separation of variables and eigenfunction expansion

Now we are looking for nonstationary solutions of the Fokker-planck equation by using the method of separation of variables along with the eigenfunction expansion. Consider that the solution $P(x, t|x', t')$ is separable in time and space given as follows:

$$P(x, t|x', t') = \phi(x, x')T(t, t'). \tag{3.24}$$

Substituting Eq. (3.24) into Eq. (3.14) we obtain

$$\frac{dT(t, t')}{dt} = -\mu T(t, t') \tag{3.25}$$

and

$$\left[ -\frac{d}{dx}D_1(x) + \frac{d^2}{dx^2}D_2(x) \right] \phi(x, x') = -\mu\phi(x, x'), \tag{3.26}$$

where $\mu$ is a separation constant and it corresponds to the eigenvalues of Eq. (3.26) with appropriate boundary conditions; the eigenvalues may be discrete or continuous or both. Eq. (3.25) gives the exponential solution:

$$T(t, t') = e^{-\mu(t-t')}. \tag{3.27}$$

The solution of the transition probability of the Fokker-Planck equation may be given by a complete set formed by the eigenvalues and eigenfunctions [5] obtained from Eq. (3.26)

$$P(x, t|x', t') = e^{\frac{\Phi(x')}{2} - \frac{\Phi(x)}{2}} \sum_n \psi_n(x)\psi_n(x')e^{-\mu_n(t-t')}, \tag{3.28}$$

where

$$\psi_n(x) = e^{\frac{\Phi(x)}{2}}\phi_n(x). \tag{3.29}$$

In the stationary state the joint probability density may be given by

$$\rho_2(x, t|x', t') = P(x, t|x', t')\rho_{st}(x'), \quad t \geq t' \tag{3.30}$$

and

$$\rho_2(x, t|x', t') = P(|x', t'|x, t)\rho_{st}(x), \quad t \leq t'. \tag{3.31}$$

### 3.2.3    *Solution for the harmonic potential*

Now we take a special case of the drift coefficient and constant diffusion coefficient given by

$$D_1(x) = \frac{F(x)}{m\gamma} \tag{3.32}$$

and

$$D_2(x) = \frac{k_B T}{m\gamma}, \tag{3.33}$$

where $\gamma$ is the friction constant. The Fokker-Planck equation (3.14) with the drift and diffusion coefficients given by Eqs. (3.32) and (3.33) is called the Smoluchowski equation which describes one-dimensional Brownian motion of a particle subjects to an external force $F(x)$.

As an example of the Fokker-Planck equation in which the solution may be given in terms of the expansion of the eigenfunctions we consider the harmonic potential given by

$$V(x) = \frac{m\omega^2 x^2}{2}. \tag{3.34}$$

Thus, the force $F(x)$ is given by

$$F(x) = -dV(x)/dx = -m\omega^2 x; \tag{3.35}$$

substituting it into Eq. (3.26) yields

$$\frac{d^2\phi(\bar{x})}{d\bar{x}^2} + \bar{x}\frac{d\phi(\bar{x})}{d\bar{x}} + (1 + \frac{\gamma\mu}{\omega^2})\phi(\bar{x}) = 0, \tag{3.36}$$

where $\bar{x} = x/x_{th}$ and $x_{th}^2 = k_B T/m\omega^2$ is the mean square position fluctuation. The process which is described by the linear drift force (3.35) with the constant diffusion coefficient is called the Ornstein-Uhlenbeck process. Substituting

$$\phi(\bar{x}) = e^{-\frac{\bar{x}^2}{2}}\bar{\phi}(\bar{x}) \tag{3.37}$$

into Eq. (3.36) we obtain

$$\frac{d^2\bar{\phi}(\bar{x})}{d\bar{x}^2} - \bar{x}\frac{d\bar{\phi}(\bar{x})}{d\bar{x}} + (1 + \frac{\gamma\mu}{\omega^2})\bar{\phi}(\bar{x}) = 0, \tag{3.38}$$

Changing the variable $\bar{x} = \sqrt{2}u$ we arrive at

$$\frac{d^2\bar{\phi}(u)}{du^2} - 2u\frac{d\bar{\phi}(u)}{du} + 2\frac{\gamma\mu}{\omega^2}\bar{\phi}(u) = 0. \tag{3.39}$$

The solution of Eq. (3.39) may be given in terms of the Hermite polynomials $H_n(u)$ [6] with $\gamma\mu_n/\omega^2 = n$, where $n = 0, 1, 2, \ldots$. Therefore, we have

$$\phi(\bar{x}) = e^{-\frac{\bar{x}^2}{2}} H_n(\frac{\bar{x}}{\sqrt{2}}). \tag{3.40}$$

The solution for the transition probability is now given by

$$P(x,t|x',t') = e^{\frac{\Phi(x')}{2} - \frac{\Phi(x)}{2}} \sum_n \psi_n(x)\psi_n(x')e^{-\mu_n(t-t')}, \tag{3.41}$$

with $\Phi(x) = V(x)/k_B T$. Substituting the solution (3.40) into (3.41) we obtain the following normalized solution:

$$P(x,t|x',t') = \frac{e^{\frac{-\bar{x}^2}{2}}}{\sqrt{2\pi x_{th}^2}} \sum_{n=0}^{\infty} \frac{1}{2^n n!} H_n\left(\frac{\bar{x}}{\sqrt{2}}\right) H_n\left(\frac{\bar{x}'}{\sqrt{2}}\right) e^{-\mu_n(t-t')}, \tag{3.42}$$

The solution (3.42) may be written in a compact form. Using the following formula:

$$\sum_{n=0}^{\infty} \frac{b^n}{n!} H_n(x) H_n(x') = \frac{1}{\sqrt{1-4b^2}} e^{\frac{4b}{\sqrt{1-4b^2}}(xx'-bx^2-bx'^2)}, \tag{3.43}$$

yields

$$P(x,t|x',t') = \frac{1}{\sqrt{2\pi x_{th}^2\left(1 - e^{-\frac{2\omega^2}{\gamma}(t-t')}\right)}} e^{-\frac{\left(\bar{x}-e^{-\frac{\omega^2}{\gamma}(t-t')}\bar{x}'\right)^2}{2\left(1-e^{-\frac{2\omega^2}{\gamma}(t-t')}\right)}}. \tag{3.44}$$

The solution (3.44) may also be obtained from the Fourier transform and the method of characteristics [7,8]. A characteristic is a curve along which a partial differential equation reduces to an ordinary differential equation [8]. The pair of the Fourier transform considered here is given by

$$P_k(k,t|x',t') = \int_{-\infty}^{\infty} e^{-ikx} P(x,t|x',t')dx \tag{3.45}$$

and

$$P(x,t|x',t') = \frac{1}{2\pi} \int_{-\infty}^{\infty} e^{ikx} P_k(k,t|x',t')dk. \tag{3.46}$$

Applying the Fourier transform to the Fokker-Planck equation with the coefficients (3.32) and (3.33) yields

$$\frac{\partial P_k(k,t|x',t')}{\partial t} + \frac{\omega^2}{\gamma} k \frac{\partial P_k(k,t|x',t')}{\partial k} = -Dk^2 P_k(k,t|x',t'), \tag{3.47}$$

where $D = k_B T/(m\gamma)$. By using the chain rule we can write Eq. (3.47) as follows:

$$\frac{dP_k(k,t|x',t')}{dt} = \frac{\partial P_k(k,t|x',t')}{\partial t} + \frac{dk}{dt}\frac{\partial P_k(k,t|x',t')}{\partial k}$$

$$= -Dk^2 P_k(k,t|x',t'), \tag{3.48}$$

with

$$\frac{dk}{dt} = \frac{\omega^2}{\gamma}k. \tag{3.49}$$

We see that the partial differential equation (3.47) is transformed into two ordinary differential equations (3.48) and (3.49). Eq. (3.49) gives a family of characteristics and its solution is given by

$$k = k' e^{\frac{\omega^2}{\gamma}(t-t')}, \tag{3.50}$$

where $k'$ corresponds to $k(t')$. By integrating Eq. (3.48) we obtain the solution for $P_k(k,t|x',t')$ along the exponential characteristics, and it is given by

$$P_k(k,t|x',t') = P'_k(k') e^{-\frac{D\gamma k'^2}{2\omega^2}\left(e^{2\frac{\omega^2}{\gamma}(t-t')}-1\right)}$$

$$= P'_k\left(ke^{-\frac{\omega^2}{\gamma}(t-t')}\right) e^{-\frac{D\gamma k^2}{2\omega^2}\left(1-e^{-2\frac{\omega^2}{\gamma}(t-t')}\right)}. \tag{3.51}$$

We now use the initial sharp condition

$$P(x,t'|x',t') = \delta(x-x') \tag{3.52}$$

to determine $P'_k(k')$. In Fourier space we have

$$P'_k(k) = e^{-ikx'}. \tag{3.53}$$

Substituting Eq. (3.53) into Eq. (3.51) yields

$$P_k(k,t|x',t') = e^{-ikx' e^{-\frac{\omega^2}{\gamma}(t-t')} - \frac{D\gamma k^2}{2\omega^2}\left(1-e^{-2\frac{\omega^2}{\gamma}(t-t')}\right)}. \tag{3.54}$$

The result (3.44) is obtained by applying the inverse Fourier transform to Eq. (3.54).

## 3.3    Solution by the method of transformation of variables

We now consider some cases of drift and diffusion coefficients which depend on temporal and spatial variables. In this case the method of separation of variables and eigenfunction expansion is not appropriate, but the solutions can be obtained by the method of the Fourier transform.

### 3.3.1 Solution for constant diffusion coefficient and the general linear force

The general linear force and constant diffusion coefficient are given by

$$D_1(x,t) = -\alpha(t)x + \beta(t) \quad \text{and} \quad D_2(x,t) = D. \tag{3.55}$$

The $\beta(t)$ function is the time dependent load force, and it has the effect of conducting the particles of the system to its direction. The Fokker-Planck equation with the coefficients (3.55) can be obtained from the following Langevin equation:

$$\frac{dx}{dt} = -\alpha(t)x + \beta(t) + \sqrt{D}L(t), \tag{3.56}$$

with the Langevin force given by

$$\langle L(t)\rangle = 0 \quad \text{and} \quad \langle L(t)L(\bar{t})\rangle = 2\delta(t - \bar{t}). \tag{3.57}$$

In this case, it is not appropriate to use the method of separation of variables and eigenfunction expansion for solving the Fokker-Planck equation such as the case of the linear force with the time independent function, but it can be solved by the method of the Fourier transform. Applying the Fourier transform to the Fokker-Planck equation with the coefficients (3.55) yields

$$\frac{\partial \rho_k(k,t)}{\partial t} = -\left[\alpha(t)k\frac{\partial}{\partial k} + i\beta(t)k + Dk^2\right]\rho_k(k,t). \tag{3.58}$$

Eq. (3.58) can be solved as follows. We seek a solution of the type

$$\rho_k(k,t) = e^{\left(\sum_{n=1}^{\infty} b_n(t)k^n\right)}. \tag{3.59}$$

Substituting Eq. (3.59) into Eq. (3.58) we obtain

$$\sum_{n=1}^{\infty}\left(\frac{db_n(t)}{dt} + n\alpha(t)b_n(t)\right)k^n + i\beta(t)k + Dk^2 = 0. \tag{3.60}$$

The solutions for the coefficients $b_n$ are given by

$$b_1(t) = b_{10}e^{-H(t,0)} - i\int_0^t d\tau\beta(\tau)e^{-H(t,\tau)}, \tag{3.61}$$

$$b_2(t) = b_{20}e^{-2H(t,0)} - D\int_0^t d\tau e^{-2H(t,\tau)} \tag{3.62}$$

and

$$b_n(t) = b_{n0}e^{-nH(t,0)} \quad \text{for} \quad n \geq 3, \tag{3.63}$$

where

$$H(t, t_1) = \int_{t_1}^{t} \alpha(\tau) d\tau. \tag{3.64}$$

Substituting the coefficients $b_n$ into Eq. (3.59) we arrive at

$$\rho_k(k, t) = e^{\left(b_{10}e^{-H(t,0)} - i\int_0^t d\tau \beta(\tau)e^{-H(t,\tau)}\right)k}$$

$$\times e^{\left(b_{20}e^{-2H(t,0)} - D\int_0^t d\tau e^{-2H(t,\tau)}\right)k^2 + \sum_{n=3}^{\infty} b_{n0}e^{-nH(t,0)}k^n}. \tag{3.65}$$

We now use the initial sharp condition

$$\rho(x, 0) = \delta(x - x_0) \tag{3.66}$$

to determine the constants $b_{n0}$. Comparing Eq. (3.53) with Eq. (3.65) we obtain

$$\rho_k(k, t) = e^{-i\left(x_0 e^{-H(t,0)} + \int_0^t d\tau \beta(\tau)e^{-H(t,\tau)}\right)k - D\int_0^t d\tau e^{-2H(t,\tau)}k^2}. \tag{3.67}$$

Applying the inverse Fourier transform to Eq. (3.67) yields

$$\rho(x, t) = \frac{1}{2\pi} \int_{-\infty}^{\infty} e^{ikx} \rho_k(k, t) dk$$

$$= \frac{1}{\sqrt{4\pi D \int_0^t d\tau e^{-2H(t,\tau)}}} e^{-\frac{\left[x - \int_0^t d\tau \beta(\tau)e^{-H(t,\tau)} - x_0 e^{-H(t,0)}\right]^2}{4D\int_0^t d\tau e^{-2H(t,\tau)}}}. \tag{3.68}$$

The first and second moments can also be obtained from the PDF (3.68), and they are given by

$$\langle x \rangle = \int_0^t d\tau \beta(\tau)e^{-H(t,\tau)} + x_0 e^{-H(t,0)} \tag{3.69}$$

and

$$\langle x^2 \rangle = 2D \int_0^t d\tau e^{-2H(t,\tau)} + \left[\int_0^t d\tau \beta(\tau)e^{-H(t,\tau)} + x_0 e^{-H(t,0)}\right]^2. \tag{3.70}$$

Besides, the variance is given by

$$\sigma_{xx} = \left\langle (x - \langle x \rangle)^2 \right\rangle = \langle x^2 \rangle - \langle x \rangle^2 = 2D \int_0^t d\tau e^{-2H(t,\tau)}. \tag{3.71}$$

Note that we can write the PDF and $n$-moment in terms of the first moment and variance as follows:

$$\rho(x, t) = \frac{1}{\sqrt{2\pi\sigma_{xx}}} e^{-\frac{[x - \langle x \rangle]^2}{2\sigma_{xx}}} \tag{3.72}$$

and

$$\langle x^n \rangle = \frac{1}{\sqrt{\pi}} \sum_{k=0}^{n} \frac{n! \left( \langle x \rangle \right)^{n-k} \left( 2\sigma_{xx} \right)^{\frac{k}{2}} \Gamma \left( \frac{1+k}{2} \right)}{k!(n-k)!}, \tag{3.73}$$

where $k$ is an even number. It should be noted that the variance is a measure of the dispersion around the mean $\langle x \rangle$, and in the case of (3.71) it does not depend on the load force. However, the presence of the load force is important to the $n$-moment.

For $\beta(t) = 0$ the PDF (3.68) reduces to [9]:

$$\rho(x,t) = \frac{1}{\sqrt{4\pi D \int_0^t d\tau e^{-2H(t,\tau)}}} e^{-\frac{\left[ x - x_0 e^{-H(t,0)} \right]^2}{4D \int_0^t d\tau e^{-2H(t,\tau)}}}. \tag{3.74}$$

For $\alpha(t) = \alpha$ the PDF (3.68) reduces to the following result:

$$\rho(x,t) = \sqrt{\frac{\alpha}{2\pi D \left( 1 - e^{-2\alpha t} \right)}} e^{-\frac{\alpha \left[ x - \int_0^t d\tau \beta(\tau) e^{-\alpha(t-\tau)} - x_0 e^{-\alpha t} \right]^2}{2D(1 - e^{-2\alpha t})}}, \tag{3.75}$$

It should be noted that the PDF (3.68) is peculiar, and it has the same solution of the time dependent linear force (3.74) with the translation of the position coordinate

$$x \;\; \rightarrow \;\; x - \int_0^t d\tau \beta(\tau) e^{-H(t,\tau)}. \tag{3.76}$$

The translation of the position coordinate can also be demonstrated from the transformation of variable $x$ which reduces the Fokker-Planck equation with the coefficients (3.55) to the equation of the linear force

$$\frac{\partial \rho(\bar{x},t)}{\partial t} = \left[ -\frac{\partial}{\partial \bar{x}} (-\alpha(t)\bar{x}) + D \frac{\partial^2}{\partial \bar{x}^2} \right] \rho(\bar{x},t), \tag{3.77}$$

with $\bar{x}$ given by the transformation (3.76).

Now the diffusion behaviors for some specific cases of the time dependent load force are investigated.

*First case.* The load force is given by a delta function, $\beta(t) = a\delta(t)$, and $\alpha(t) = \alpha$. In this case the PDF, the first and second moments are described by

$$\rho(x,t) = \sqrt{\frac{\alpha}{2\pi D \left( 1 - e^{-2\alpha t} \right)}} e^{-\frac{\alpha \left[ x - (a+x_0) e^{-\alpha t} \right]^2}{2D(1 - e^{-2\alpha t})}}, \tag{3.78}$$

$$\langle x \rangle = (a + x_0) e^{-\alpha t} \tag{3.79}$$

and

$$\langle x^2 \rangle = \frac{D \left(1 - e^{-2\alpha t}\right)}{\alpha} + (a + x_0)^2 \, e^{-2\alpha t}. \tag{3.80}$$

Note that for $a = -x_0$ the peak of the PDF is dislocated to the origin for any time. The thermal equilibrium $\langle x^2 \rangle = D/\alpha$ is reached for $\alpha > 0$ and $t \to \infty$.

*Second case.* The load force is a constant, $\beta(t) = a$, and $\alpha(t) = \alpha$. In this case the PDF, the first and second moments are described by

$$\rho(x,t) = \sqrt{\frac{\alpha}{2\pi D \left(1 - e^{-2\alpha t}\right)}} e^{-\frac{\alpha \left[x - \left(\frac{a}{\alpha}(e^{\alpha t} - 1) + x_0\right) e^{-\alpha t}\right]^2}{2D \left(1 - e^{-2\alpha t}\right)}}, \tag{3.81}$$

$$\langle x \rangle = \left(\frac{a}{\alpha} \left(e^{\alpha t} - 1\right) + x_0\right) e^{-\alpha t} \tag{3.82}$$

and

$$\langle x^2 \rangle = \frac{D \left(1 - e^{-2\alpha t}\right)}{\alpha} + \left(\frac{a}{\alpha} \left(e^{\alpha t} - 1\right) + x_0\right)^2 e^{-2\alpha t}. \tag{3.83}$$

One can see that the stationary state is preserved for the harmonic potential ($\alpha > 0$), and it is given by

$$\rho_{st}(x) = \sqrt{\frac{\alpha}{2\pi D}} e^{-\frac{\alpha \left(x - \frac{a}{\alpha}\right)^2}{2D}}; \tag{3.84}$$

the peak of the PDF is dislocated to the position $a/\alpha$ for $t \gg 1$. For $\alpha > 0$, the first moment $\langle x \rangle$, without the load force, decays to zero due to the confining potential; however, the first moment, with the presence of the load force, shows a net drift in a direction determined by the constant load force, and it decays to $a/\alpha$ exponentially. This means that the particles of the system are preferably conducted to the direction of the constant load force. In this case the thermal equilibrium is given by $\langle x^2 \rangle = D/\alpha + a^2/\alpha^2$.

*Third case.* The load force is given by $\beta(t) = bt$, and $\alpha(t) = \alpha$. In this case the PDF, the first and second moments are described by

$$\rho(x,t) = \sqrt{\frac{\alpha}{2\pi D \left(1 - e^{-2\alpha t}\right)}} e^{-\frac{\alpha \left[x - b\left(\frac{t}{\alpha} - \frac{1}{\alpha^2}\right) - \left(\frac{b}{\alpha^2} + x_0\right) e^{-\alpha t}\right]^2}{2D(1 - e^{-2\alpha t})}}, \tag{3.85}$$

$$\langle x \rangle = \left(\frac{b}{\alpha} \left(t - \frac{1}{\alpha}\right) + \left(\frac{b}{\alpha^2} + x_0\right) e^{-\alpha t}\right) \tag{3.86}$$

and

$$\langle x^2 \rangle = \frac{D \left(1 - e^{-2\alpha t}\right)}{\alpha} + \left(\frac{b}{\alpha} \left(t - \frac{1}{\alpha}\right) e^{\alpha t} + \frac{b}{\alpha^2} + x_0\right)^2 e^{-2\alpha t}. \tag{3.87}$$

The peak of the PDF is dislocated with the time increase.

*Fourth case.* The load force is given by $\beta(t) = a_1 e^{-a_2 t}$, and $\alpha(t) = \alpha$. In this case the PDF, the first and second moments are described by

$$\rho(x,t) = \sqrt{\frac{\alpha}{2\pi D \left(1 - e^{-2\alpha t}\right)}} e^{-\frac{\alpha\left[x - \left(\frac{a_1}{\alpha - a_2}\left(e^{(\alpha - a_2)t} - 1\right) + x_0\right)e^{-\alpha t}\right]^2}{2D\left(1 - e^{-2\alpha t}\right)}}, \qquad (3.88)$$

$$\langle x \rangle = \left(\frac{a_1}{\alpha - a_2}\left(e^{(\alpha - a_2)t} - 1\right) + x_0\right)e^{-\alpha t} \qquad (3.89)$$

and

$$\langle x^2 \rangle = \frac{D\left(1 - e^{-2\alpha t}\right)}{\alpha} + \left(\frac{a_1}{\alpha - a_2}\left(e^{(\alpha - a_2)t} - 1\right) + x_0\right)^2 e^{-2\alpha t}. \qquad (3.90)$$

In particular, for $a_2 = a\alpha$, $a_1 = (1 - a)\alpha x_0$ and $a \neq 1$ the PDF (3.44) of the linear force is reproduced with the decay of the initial position $x_0$ given by $e^{-a\alpha t}$. For $\alpha > 0$ and $a_2 > 0$ the first moment $\langle x \rangle$ decays to zero and the thermal equilibrium is given by $\langle x^2 \rangle = D/\alpha$.

### 3.3.2 *Solution for time dependent drift coefficient and variable diffusion coefficient*

We now consider the Fokker-Planck equation in the Stratonovich prescription ($\lambda = 1/2$) described by Eq. (3.5), and we assume that the deterministic drift $h_1(x,t)$ and multiplicative noise term $h_2(x,t)$ are separable in time and space given by

$$h_1(x,t) = T(t)\,x \qquad (3.91)$$

and

$$h_2(x,t) = ax. \qquad (3.92)$$

From Eq. (3.5) we have

$$\frac{\partial \rho(x,t)}{\partial t} = -T(t)\frac{\partial}{\partial x}\left[x\rho(x,t)\right] + a^2\frac{\partial}{\partial x}\left[x\frac{\partial}{\partial x}\left(x\rho(x,t)\right)\right]. \qquad (3.93)$$

Now we introduce the following transformations:

$$\bar{\rho}(x,t) = a|x|\rho(x,t) \qquad (3.94)$$

and

$$y = \frac{\ln|x|}{a}. \qquad (3.95)$$

Applying the transformations (3.94) and (3.95) to Eq. (3.93) yields

$$\frac{\partial \bar{\rho}(y,t)}{\partial t} = -\frac{T(t)}{a}\frac{\partial \bar{\rho}(y,t)}{\partial y} + \frac{\partial^2 \bar{\rho}(y,t)}{\partial y^2}. \tag{3.96}$$

The Fourier transform of Eq. (3.96) is given by

$$\frac{\partial \bar{\rho}_k(k,t)}{\partial t} = -\left[\frac{ikT(t)}{a} + k^2\right]\bar{\rho}_k(k,t), \tag{3.97}$$

where $\bar{\rho}_k(k,t)$ is the Fourier transform of $\bar{\rho}(y,t)$. We now seek a solution of the type

$$\bar{\rho}_k(k,t) = e^{\left(\sum_{n=1}^{\infty} b_n(t)k^n\right)}. \tag{3.98}$$

Substituting Eq. (3.98) into Eq. (3.97) we have

$$\sum_{n=1}^{\infty}\frac{db_n(t)}{dt}k^n + \frac{ikT(t)}{a} + k^2 = 0. \tag{3.99}$$

The solutions for the coefficients $b_n$ are given by

$$b_1(t) = -\frac{i}{a}H(t,0) + b_{10}, \tag{3.100}$$

$$b_2(t) = -t + b_{20} \tag{3.101}$$

and

$$b_n(t) = 0 \quad \text{for} \quad n \geq 3, \tag{3.102}$$

where $H(t,t') = \int_{t'}^{t} T(\tau)d\tau$. Substituting the coefficients $b_n$ into Eq. (3.98) we arrive at

$$\bar{\rho}_k(k,t) = e^{\left(b_{10} - \frac{i}{a}H(t,0)\right)k + (b_{20}-t)k^2}. \tag{3.103}$$

We now use the initial condition

$$\rho(x,0) = \delta(x - x_0) \tag{3.104}$$

to determine the constants $b_{10}$ and $b_{20}$; in Fourier space we have

$$\bar{\rho}_k(k,0) = e^{-iky_0}. \tag{3.105}$$

Comparing Eq. (3.105) with Eq. (3.103) we obtain

$$\bar{\rho}_k(k,t) = e^{-i\left(y_0 + \frac{H(t,0)}{a}\right)k - tk^2}. \tag{3.106}$$

Applying the inverse Fourier transform to Eq. (3.106) yields

$$\bar{\rho}(y,t) = \frac{1}{2\pi}\int_{-\infty}^{\infty} e^{iky}\bar{\rho}_k(k,t)dk = \frac{e^{-\frac{B^2}{4t}}}{2\pi}\int_{-\infty}^{\infty} e^{-t\left(k - \frac{iB}{2t}\right)^2}dk, \tag{3.107}$$

where $B = y - y_0 - \frac{H(t,0)}{a}$. The last integral of Eq. (3.107) gives $\sqrt{\pi/t}$, thus we arrive at

$$\bar{\rho}(y,t) = \frac{e^{-\frac{\left(y - y_0 - \frac{H(t,0)}{a}\right)^2}{4t}}}{\sqrt{4\pi t}}. \tag{3.108}$$

In terms of the variable $x$ and $\rho(x,t)$ we have the following normalized PDF:

$$\rho(x,t) = \frac{e^{-\frac{\left(\ln\left|\frac{x}{x_0}\right| - H(t,0)\right)^2}{4a^2 t}}}{4\sqrt{\pi a^2 t}|x|}. \tag{3.109}$$

Note that the PDF (3.109) does not diverge for $x \to 0$. To see that, we write

$$\frac{e^{-A(\ln|x|)^2}}{|x|} = e^{-\ln|x| - A(\ln|x|)^2}, \tag{3.110}$$

and considering that $t \neq 0$ and $H(t,0)$ is bounded. The second-order term in the exponential is the dominant term for $x \to 0$, therefore, the PDF tends to zero. In Fig. 3.2 we show the PDF vs position in different time points. It is interesting to note that the multiplicative noise term has the effect of splitting the peak of the PDF in two peaks in $x$ coordinate for a fixed time.

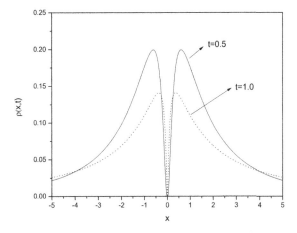

Fig. 3.2 Plots of the PDF, Eq. (3.109), vs position in different time points, for $H(t,0) = t$, and $a = x_0 = 1$.

## 3.4 Langevin equation with multiplicative white noise: Transformation of diffusion processes into the Wiener process in different prescriptions

In this section, we consider a Langevin equation with multiplicative white noise and its corresponding Fokker-Planck equation. From the Fokker-Planck equation a transformation into the Wiener process is provided for different orders of prescription in discretization rules for the stochastic integrals. A few applications are also discussed.

### 3.4.1 Deterministic drift $h_1(x, t)$ and multiplicative noise term $h_2(x, t)$ are separable in time and space

We first consider the Fokker-Planck equation in the Stratonovich prescription ($\lambda = 1/2$) given by Eq. (3.5), and we assume that the deterministic drift $h_1(x, t)$ and multiplicative noise term $h_2(x, t)$ are separable in time and space given by

$$h_1(x, t) = T_1(t) D(x) \tag{3.111}$$

and

$$h_2(x, t) = T_2(t)D(x). \tag{3.112}$$

From Eq. (3.5) we obtain

$$\frac{\partial \rho(x, t)}{\partial t} = -T_1(t) \frac{\partial}{\partial x} [D(x)\rho(x, t)]$$

$$+ T_2^2(t)\frac{\partial}{\partial x} \left[ D(x)\frac{\partial D(x)\rho(x, t)}{\partial x} \right]. \tag{3.113}$$

By suitable transformations of variables we can show that Eq. (3.113) can be reduced to a constant-diffusion equation without the drift coefficient term. To do so, we take the following transformations:

$$\bar{\rho}(x, t) = D(x)\rho(x, t) , \tag{3.114}$$

$$\bar{t} = \bar{t}(t) \quad \text{and} \quad \bar{x} = \bar{x}(x, t). \tag{3.115}$$

Applying the transformations (3.114) and (3.115) to Eq. (3.113) yields

$$\frac{\partial \bar{\rho}}{\partial \bar{t}}\frac{d\bar{t}}{dt} = - \left[ \frac{\partial \bar{x}}{\partial t} + T_1(t)D(x)\frac{\partial \bar{x}}{\partial x} \right] \frac{\partial \bar{\rho}}{\partial \bar{x}}$$

$$+ T_2^2(t)D(x)\frac{\partial \bar{x}}{\partial x} \frac{\partial}{\partial \bar{x}} \left[ D(x)\frac{\partial \bar{x}}{\partial x} \frac{\partial \bar{\rho}}{\partial \bar{x}} \right]. \tag{3.116}$$

Now we take

$$\frac{d\bar{t}}{dt} = T_2^2(t) \tag{3.117}$$

and

$$\bar{x} = \int \frac{dx}{D(x)} - \int dt T_1(t) + A, \tag{3.118}$$

where $A$ is a constant, then Eq. (3.116) reduces to

$$\frac{\partial \bar{\rho}(\bar{t}, \bar{x})}{\partial \bar{t}} = \frac{\partial^2 \bar{\rho}(\bar{t}, \bar{x})}{\partial \bar{x}^2}. \tag{3.119}$$

Eqs. (3.117) and (3.118) give the time and space scaling factors which transform Eq. (3.113) into the ordinary diffusion equation or the Wiener process (3.119). Eq. (3.119) can be solved and the solution with a natural boundary condition is given by

$$\bar{\rho}(\bar{t}, \bar{x}) = C \frac{\exp\left[-\frac{\bar{x}^2}{4\bar{t}}\right]}{\sqrt{\bar{t}}} \tag{3.120}$$

where $C$ is a normalization factor. Eqs. (3.114) and (3.120) show that the time-dependent coefficients $T_1(t)$ and $T_2(t)$ do not change the Gaussian form, however the coefficient $D(x)$ can produce different forms for the distribution $\rho(x,t)$ [3]. We note that for $D(x) = \sqrt{D}$, $T_1(t) = 0$ and $T_2(t) = 1$ the Wiener process is recovered.

In order to investigate some details of the solution (3.120) we take $D(x) = \sqrt{D}\,|x|^{-\frac{\theta}{2}}$. From Eqs. (3.114), (3.118) and (3.120), with $A = 0$ and $\theta \neq -2$, yields

$$\rho(x,t) = \begin{cases} \dfrac{(-x)^{\frac{\theta}{2}}}{\sqrt{4\pi D\bar{t}(t)}} \exp\left[ -\dfrac{\left((-x)^{\frac{2+\theta}{2}} + \frac{\sqrt{D}(2+\theta)}{2} H(t)\right)^2}{D(2+\theta)^2 \bar{t}(t)} \right], & x < 0 \\[4mm] \dfrac{x^{\frac{\theta}{2}}}{\sqrt{4\pi D\bar{t}(t)}} \exp\left[ -\dfrac{\left(x^{\frac{2+\theta}{2}} - \frac{\sqrt{D}(2+\theta)}{2} H(t)\right)^2}{D(2+\theta)^2 \bar{t}(t)} \right], & x > 0 \end{cases}, \tag{3.121}$$

where $H(t) = \int dt T_1(t)$. Eq. (3.121) shows that the drift term produces an asymmetric PDF with respect to the coordinate $x$. In this case, the drift term $T_1(t)$ gives the duration of this asymmetry. For $T_1(t) = 0$ the PDF (3.121) recovers the symmetric PDF which is given by

$$\rho(x,\bar{t}) = \frac{|x|^{\frac{\theta}{2}}}{\sqrt{4\pi D\bar{t}(t)}} \exp\left[ -\frac{|x|^{2+\theta}}{D(2+\theta)^2 \bar{t}(t)} \right]. \tag{3.122}$$

In Fig. 3.3 we show the asymmetric PDF (3.121) for $t = 0.1$. The asymmetry of the PDF with $\theta = -0.1$ is more pronounced than the PDF with $\theta = -0.4$. From Eq. (3.121) we obtain the following second moment:

$$\left\langle x^2(t) \right\rangle = \frac{\Gamma\left(\frac{6+\theta}{2(2+\theta)}\right)}{\sqrt{\pi}} \times$$

$$\left[ D\left(2+\theta\right)^2 \bar{t}\left(t\right) \right]^{\frac{2}{2+\theta}} e^{-\frac{H^2(t)}{\bar{t}(t)}} \,_1F_1\left(\frac{6+\theta}{2\left(2+\theta\right)}, \frac{1}{2}, \frac{H^2(t)}{\bar{t}\left(t\right)}\right), \qquad (3.123)$$

where $\,_1F_1\left(\frac{6+\theta}{2(2+\theta)}, \frac{1}{2}, \frac{H^2(t)}{\bar{t}(t)}\right)$ is the Kummer confluent hypergeometric function [10].

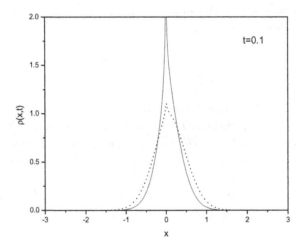

Fig. 3.3   Plots of the PDF (3.121) for $D = 1$, $\bar{t}(t) = t$ and $H(t) = t$. The solid line corresponds to $\theta = -0.4$, whereas the dotted line corresponds to $\theta = -0.1$.

In particular, for $A = T_1 = 0$ and $\theta = -2$ the PDF is given by

$$\rho(x, \bar{t}) = \frac{1}{4\sqrt{\pi D \bar{t}\left(t\right)}|x|} \exp\left[-\frac{(\ln|x|)^2}{4D\bar{t}\left(t\right)}\right], \qquad (3.124)$$

and it is the log-normal distribution; its second moment gives a simple exponential function

$$\left\langle x^2(t) \right\rangle = e^{4D\bar{t}}. \qquad (3.125)$$

We should note that the solutions (3.114) and (3.120) can adequately work for $D(x)$ positive. For $D(x)$ containing negative values we should modify and take $\bar{\rho}(x,t) = -D(x)\rho(x,t)$ for $D(x)$ negative. For example, let us consider $D(x) = x$. Then we take

$$\bar{\rho}(x,t) = -x\rho(x,t),\ x < 0 \tag{3.126}$$

and

$$\bar{\rho}(x,t) = x\rho(x,t),\ x > 0\ . \tag{3.127}$$

From Eq. (3.118) we obtain

$$\bar{x} = \ln|x| - H(t) - \ln|x_0|$$

and

$$\rho(x,t) = \frac{\exp\left[-\frac{(\ln|x|-H(t)-\ln|x_0|)^2}{4\bar{t}(t)}\right]}{4\sqrt{\pi\bar{t}(t)}\,|x|}\ . \tag{3.128}$$

This is the log-normal distribution. The distribution (3.128) is the same as the one given in Ref. [11] for $\bar{t}(t) = t$, which has been obtained from a different method.

One can see that when a multiplicative noise term is introduced into the simple Langevin equation (2.1), even separable in time and space, the system can exhibit complex behaviors and a rich variety of processes.

### 3.4.2   *General drift $h_1(x,t)$ and multiplicative noise term $h_2(x,t)$*

Now, we consider the following transformations [12, 13]:

$$\bar{x} = \psi(x,t)\ ,\ \bar{t} = \phi(t) \tag{3.129}$$

and

$$\rho(x,t) = \bar{J}\bar{\rho}(\bar{x},\bar{t}) = \frac{1}{J}\bar{\rho}(\bar{x},\bar{t}) = \frac{\partial\psi(x,t)}{\partial x}\bar{\rho}(\bar{x},\bar{t})\ , \tag{3.130}$$

where $J$ is the Jacobian of the transformation given by

$$J = \frac{\partial x}{\partial\bar{x}} = \frac{1}{\bar{J}} = \frac{1}{\frac{\partial\bar{x}}{\partial x}}\ . \tag{3.131}$$

The transformations (3.129) and (3.130) permit us to change Eq. (3.1) into the Fokker-Planck equation with a constant diffusion coefficient, namely, the Wiener process (3.119), and its solution is given by

$$\bar{\rho}(\bar{x},\bar{t}) = \frac{1}{\sqrt{4\pi\bar{t}}}e^{-\frac{\bar{x}^2}{4\bar{t}}}\ . \tag{3.132}$$

In order to obtain the explicit forms of $\psi(x,t)$ and $\phi(t)$, the transformations (3.129) and (3.130) are applied to Eq. (3.1). Using the chain rule yields

$$\frac{\partial}{\partial x} = \frac{1}{J} \frac{\partial}{\partial \overline{x}} \left( \frac{\partial \overline{x}}{\partial x} J \right) , \tag{3.133}$$

$$\frac{\partial^2}{\partial x^2} = \frac{1}{J} \frac{\partial^2}{\partial \overline{x}^2} \left[ \left( \frac{\partial \overline{x}}{\partial x} \right)^2 J \right] - \frac{1}{J} \frac{\partial}{\partial \overline{x}} \left( \frac{\partial^2 \overline{x}}{\partial x^2} J \right) \tag{3.134}$$

and

$$\frac{\partial}{\partial t} = \frac{1}{J} \left( \frac{\partial \overline{t}}{\partial t} \right) \frac{\partial}{\partial \overline{t}} J + \frac{1}{J} \frac{\partial}{\partial \overline{x}} \left( \frac{\partial \overline{x}}{\partial t} J \right) . \tag{3.135}$$

Substituting Eq. (3.130) and Eqs. (3.133)-(3.135) into Eq. (3.1) we arrive at

$$\frac{\partial \overline{p}(\overline{x}, \overline{t})}{\partial \overline{t}} = -\frac{\partial}{\partial \overline{x}} \left[ \overline{D}_1(x,t) \overline{p}(\overline{x}, \overline{t}) \right] + \frac{\partial^2}{\partial \overline{x}^2} \left[ \overline{D}_2(x,t) \overline{p}(\overline{x}, \overline{t}) \right] , \tag{3.136}$$

where

$$\overline{D}_1(x,t) = \frac{1}{\phi'(t)} \times$$

$$\left[ \frac{\partial \psi(x,t)}{\partial t} + D_1(x,t) \frac{\partial \psi(x,t)}{\partial x} + D_2(x,t) \frac{\partial^2 \psi(x,t)}{\partial x^2} \right] \tag{3.137}$$

and

$$\overline{D}_2(x,t) = \frac{D_2(x,t)}{\phi'(t)} \left[ \frac{\partial \psi(x,t)}{\partial x} \right]^2 ; \tag{3.138}$$

where the prime indicates the derivative with respect to $t$. In order to obtain the Wiener process (3.119) we put the coefficients of Eq. (3.136) equal to $\overline{D}_1(x,t) = 0$ and $\overline{D}_2(x,t) = 1$. Setting $\overline{D}_2(x,t) = 1$ and making use of Eq. (3.138) gives

$$\psi(x,t) = \sqrt{\phi'(t)} \int_z^x \frac{dy}{\sqrt{D_2(y,t)}} + \chi(t) , \tag{3.139}$$

where $\chi(t)$ is a function of $t$. Now, setting $\overline{D}_1(x,t) = 0$ and making use of Eqs. (3.137) and (3.139) yields

$$\frac{h_1(x,t)}{h_2(x,t)} = (1 - 2\lambda) \frac{\partial h_2(x,t)}{\partial x}$$

$$+ \frac{C_1(t)}{2} + \int_z^x \frac{C_2(t)h_2(y,t) + 2h_2'(y,t)}{2h_2^2(y,t)} dy, \tag{3.140}$$

where

$$C_1(t) = -\frac{2\chi'(t)}{\sqrt{\phi'(t)}} \tag{3.141}$$

and

$$C_2(t) = -\frac{\phi''(t)}{\phi'(t)} . \tag{3.142}$$

Notice that the identity (3.140) is valid for any prescription.

From Eqs. (3.139), (3.141) and (3.142) one can obtain the explicit forms of $\psi(x,t)$ and $\phi(t)$ given by

$$\phi(t) = \phi'(t_0) \int_{t_1}^t e^{-\int_{t_0}^\tau C_2(\tau)d\tau} + \phi(t_1) \tag{3.143}$$

and

$$\psi(x,t) = \chi(t_2) + \sqrt{\phi'(t_0)}$$

$$\times \left[ e^{-\int_{t_0}^t \frac{C_2(\tau)}{2} d\tau} \int_z^x \frac{dy}{\sqrt{D_2(y,t)}} - \int_{t_2}^t \frac{C_1(\tau)}{2} e^{-\int_{t_0}^\tau \frac{C_2(\tau_1)}{2} d\tau_1} d\tau \right], \tag{3.144}$$

where the functions $C_1(t)$ and $C_2(t)$ are determined by Eq. (3.140).

### 3.4.3 *Applications*

As applications of the above formulas we consider the following systems:

*Case 1.* We consider the following drift and diffusion coefficients:

$$h_1(x,t) = [\alpha - C(t)]x - \beta x \ln x \tag{3.145}$$

and

$$h_2(x,t) = \sigma x . \tag{3.146}$$

This is a Gompertz-type model [14], which has been used to model the dynamics of a tumor population represented by $x(t)$, in the Ito prescription. The parameters $\alpha$, $\beta$ and $\sigma$ are positive constants representing the growth and death rates and the amplitude of the fluctuation, respectively. The function $C(t)$ represents the effects of a therapy. The PDF of this system can be exactly determined for any prescription parameter. The transformation variables $\phi(t)$ and $\psi(x,t)$ are obtained by using Eqs. (3.140)-(3.146), and they are given by

$$\phi(t) = \phi(t_1) + \frac{\phi'(t_0) e^{2\beta t_0}}{2\beta} \left[ e^{2\beta t} - e^{2\beta t_1} \right] \tag{3.147}$$

and

$$\psi(x,t) = \frac{\sqrt{\phi'(t_0)}}{\sigma e^{\beta t_0}} \left\{ e^{\beta t} \ln x - e^{\beta t_2} \ln z \right.$$

$$\left. - \left[ \alpha - (1 - 2\lambda) \sigma^2 \right] \frac{e^{\beta t} - e^{\beta t_2}}{\beta} + \int_{t_2}^{t} C(\tau) e^{\beta \tau} d\tau \right\} + \chi(t_2) \ . \tag{3.148}$$

Thus, the PDF is given by

$$\rho(x,t) = \frac{1}{\sqrt{2\pi \sigma_x^2(t)} x} e^{-\frac{\left[ \ln x - M_x^2(t) \right]^2}{2\sigma_x^2(t)}} \ , \tag{3.149}$$

where

$$\sigma_x^2(t) = \frac{\sigma^2}{\beta} \left[ 1 - e^{-2\beta(t-\tau)} \right] \tag{3.150}$$

and

$$M_x^2(t) = e^{-\beta(t-\tau)} \ln z +$$

$$\left[ \alpha - (1 - 2\lambda) \sigma^2 \right] \frac{1 - e^{-\beta(t-\tau)}}{\beta} - e^{-\beta t} \int_{\tau}^{t} C(\tau_1) e^{\beta \tau_1} d\tau_1 \ . \tag{3.151}$$

As one can see, the PDF presents a log-normal distribution for all of the prescriptions. Moreover, the coefficient $\left[ \alpha - (1 - 2\lambda) \sigma^2 \right]$ shows the PDF of a prescription may describe the same behavior of another prescription by changing the growth rate $\alpha$. It means that, in this system, the different orders of prescription may be associated with the growth rate.

*Case 2.* Now, we consider the models of population growth [15] which are frequently described by nonlinear differential equations without the independent variable (time) explicitly. For instance, the classical Verhulst logistic equation is a simple nonlinear model of population growth and it has been employed as a starting point to formulate various generalized models [16–23]; this logistic equation has been successfully used to model many laboratory populations such as yeast growth in laboratory cultures, growth of the Tasmanian and South Australian sheep populations [17] and self-organization at the macromolecular level [24, 25]. Besides, the parameters involved in these models are subject to fluctuations and various types of noises which may affect the replication processes. We consider the deterministic drift $h_1(x,t)$ and the multiplicative noise term $h_2(x,t)$ given by

$$h_1(x,t) = rD(x),$$ (3.152)

$$h_2(x,t) = bD(x)$$ (3.153)

and

$$D(x) = x\frac{(K^\nu - x^\nu)}{\mu K^\nu - (\mu - \nu)x^\nu},$$ (3.154)

where $x(t)$ is the number of population alive at time $t$, $\mu$ and $\nu$ are positive real parameters, $r$ is the intrinsic growth rate, and $K$ is the carrying capacity. The model described by Eqs. (3.152)-(3.154) contains the classical growth models such as the Verhulst logistic model ($\mu = 1$ and $\nu = 1$), Gompertz model ($\mu = 0$ and $\nu \to 0$), Shoener model ($\mu = 0$ and $\nu = 1$), Richards model ($\mu = 0$ and $0 \leq \nu < \infty$) and Smith model ($0 \leq \mu < \infty$ and $\nu = 1$).

For the system described by Eqs. (2.1) and (3.152) without the influence of the noise ($h_2(x) = 0$)

$$\frac{dx}{dt} = r\frac{x\left[1 - \left(\frac{x}{K}\right)^\nu\right]}{\mu\left[1 - \left(1 - \frac{\nu}{\mu}\right)\left(\frac{x}{K}\right)^\nu\right]},$$ (3.155)

the solution can be determined implicitly and it is given by

$$rt = \ln\left(\frac{\left(\frac{x}{K}\right)^\mu\left(1 - \left(\frac{x_0}{K}\right)^\nu\right)}{\left(\frac{x_0}{K}\right)^\mu\left(1 - \left(\frac{x}{K}\right)^\nu\right)}\right),$$ (3.156)

where $x_0 = x(t = 0)$ is the initial value. Fig. 3.4 shows the evolution of population for different models; they exhibit sigmoidal shapes.

For the model described by the coefficients (3.152)-(3.154) it seems that the relation (3.140) can not hold for all of the prescriptions. In this case, we only consider the Stratonovich prescription, and the solution for the PDF is given by

$$\rho(x,t) = \frac{C\exp\left(-\frac{(H(x)-rt)^2}{4Dt}\right)}{\sqrt{DtD(x)}},$$ (3.157)

where $C$ is the normalization constant. The normalization is taken by

$$\int_0^K dx\rho(x,t) = \frac{C}{\sqrt{Dt}}\int_0^K dx\frac{\exp\left(-\frac{(H(x)-rt)^2}{4Dt}\right)}{D(x)} = 1.$$ (3.158)

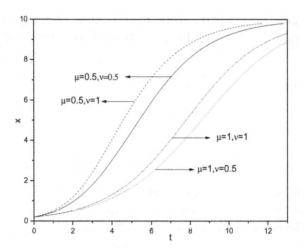

Fig. 3.4   Plots of the evolution of population for different growth models described by Eq. (3.156), in arbitrary units. The parameter values are given by $r = 0.5$, $x_0 = 0.2$ and $K = 10$.

Thus, we have

$$C = \frac{\sqrt{Dt}}{\int_0^K dx \frac{\exp\left(-\frac{(H(x)-rt)^2}{4Dt}\right)}{D(x)}} = \frac{\sqrt{Dt}}{\int_{H(0)-rt}^{H(K)-rt} du \frac{\exp\left(-\frac{u^2}{4Dt}\right)}{D(x)}} \qquad (3.159)$$

and

$$\rho(x,t) = \frac{\exp\left(-\frac{(H(x)-rt)^2}{4Dt}\right)}{\int_{H(0)-rt}^{H(K)-rt} du \exp\left(-\frac{u^2}{4Dt}\right) D(x)}, \qquad (3.160)$$

where $D = b^2$,

$$H(x) = \ln\left[\frac{\left(\frac{x}{K}\right)^\mu \left(1 - \left(\frac{x_0}{K}\right)^\nu\right)}{\left(1 - \left(\frac{x}{K}\right)^\nu\right) \left(\frac{x_0}{K}\right)^\mu}\right], \qquad (3.161)$$

$x_0 = x(t = 0)$ and the quantity $x(t)$ is limited to the value of $K$.

The integral of Eq. (3.159) can be calculated exactly for a given time $t$. One first considers for any $\mu$ and $\nu$ except for $\mu = 0$ and $\nu > 0$. In this case, for $x \to 0$, $H(x)$ gives $H(x \to 0) \to -\infty$; whereas for $x \to K$, it gives $H(x \to K) \to \infty$. Thus, the integral yields

$$\int_{H(0)-rt}^{H(K)-rt} du \exp\left(-\frac{u^2}{4Dt}\right) = \int_{-\infty}^{\infty} du \exp\left(-\frac{u^2}{4Dt}\right) = 2\sqrt{\pi Dt}. \quad (3.162)$$

The normalization constant and the PDF (3.160) are now given by

$$C = \frac{1}{2\sqrt{\pi}} \tag{3.163}$$

and

$$\rho(x,t) = \frac{\exp\left(-\frac{(H(x)-rt)^2}{4Dt}\right)}{2\sqrt{\pi Dt}D(x)}. \tag{3.164}$$

Notice that the PDF (3.164) describes a generalized log-normal distribution. The n-moment is obtained from the following expression:

$$\langle x^n(t) \rangle = \int_0^K dx x^n \rho(x,t) \;. \tag{3.165}$$

Substituting Eq. (3.164) into (3.165) yields

$$\langle x^n(t) \rangle = \int_0^K dx x^n \frac{\exp\left(-\frac{(H(x)-rt)^2}{4Dt}\right)}{2\sqrt{\pi Dt}D(x)} \;. \tag{3.166}$$

For the Verhulst logistic model ($\mu = \nu = 1$) the PDF and n-moment are given by

$$\rho(x,t) = \frac{\exp\left(-\frac{\left(\ln\left(\frac{x}{1-x/K}\frac{1-x_0/K}{x_0}\right)-rt\right)^2}{4Dt}\right)}{2\sqrt{\pi Dt}x\left(1-x/K\right)} \tag{3.167}$$

and

$$\langle x^n(t) \rangle = \frac{K^n}{\sqrt{\pi}} \int_{-\infty}^{\infty} dz \frac{\exp\left(-z^2\right)}{\left[1 + B\exp\left(-2\sqrt{Dt}z - rt\right)\right]^n} \;, \tag{3.168}$$

where $B = (K - x_0)/x_0$. For $n = 1$ and $K = 1$, the solution (3.168) is exactly the same as the one given in Ref. [25] which has been obtained by another method; in that work the authors did not obtain the PDF. Fig. 3.5 shows the behavior of the PDF $\rho(x,t)$. The PDF presents monomodal distribution for $D = 0.5$, but it presents transition from a monomodal to bimodal distribution for $D = 1$. Fig. 3.6 shows the mean value $\langle x(t) \rangle$ for two different values of $D$.

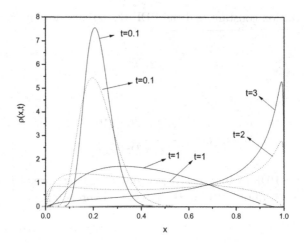

Fig. 3.5    Plots of the PDF, Eq. (3.167), versus $x$ for different time points. The values of the parameters are: $r = 1$, $x_0 = 0.2$ and $K = 1$. The solid lines correspond to $D = 0.5$ and the dotted lines correspond to $D = 1$.

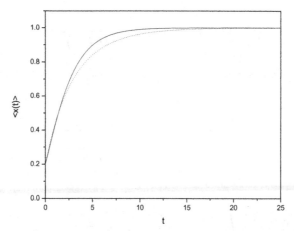

Fig. 3.6    Plots of the mean value $\langle x(t) \rangle$ versus time using Eq. (3.168) for the following parameter values: $r = 1$, $x_0 = 0.2$ and $K = 1$. The solid line corresponds to $D = 1$ and the dotted line corresponds to $D = 1.5$.

In particular, for $\nu \to 0$, the PDF (3.160) yields

$$\rho(x,t) = \left(1 - \mu \ln\left(\frac{K}{x}\right)\right) \frac{\exp\left(-\frac{\left(\ln\left|\frac{\ln\left(\frac{x}{K}\right)}{x^\mu}\right| - \ln\left|\frac{\ln\left(\frac{x_0}{K}\right)}{x_0^\mu}\right| + rt\right)^2}{4Dt}\right)}{2\sqrt{\pi D t} x \ln\left(\frac{K}{x}\right)}. \quad (3.169)$$

As one can see, the PDF (3.169) reduces to the Gompertz model for $\mu = 0$ which is given by

$$\rho(x,t) = \frac{\exp\left(-\frac{\left(\ln\left|\frac{\ln\left(\frac{x}{K}\right)}{\ln\left(\frac{x_0}{K}\right)}\right| + rt\right)^2}{4Dt}\right)}{2\sqrt{\pi Dt}x\ln\left(\frac{K}{x}\right)}. \tag{3.170}$$

This last result is the same as the one obtained in Ref. [26].

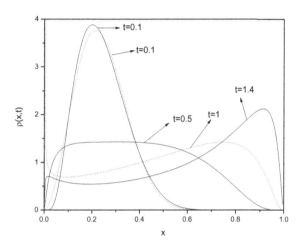

Fig. 3.7   Plots of the PDF, Eqs. (3.164) and (3.170), versus $x$ for the generalized Verhulst logistic model (3.154). The parameter values are: $r = 1$, $x_0 = 0.2$, $K = 1$ and $D = 0.5$. The solid lines correspond to $\mu = 0.5$ and $\nu = 1.5$, whereas the dotted lines correspond to the Gompertz model.

Fig. 3.7 shows the behavior of the PDF given by Eq. (3.164) for the generalized Verhulst logistic model Eqs. (3.152)-(3.154). The PDF presents transition from a monomodal to bimodal distribution for $D = 0.5$, $\mu = 0.5$ and $\nu = 1.5$. Fig. 3.8 shows the second moment for different models; the curves can be fitted by the power-law function at intermediate times which describe subdiffusive regimes according to Eq. (1.1).

For $\mu = 0$ and $\nu > 0$, the integral of Eq. (3.160) yields

$$\int_{H(0)-rt}^{H(K)-rt} du \exp\left(-\frac{u^2}{4Dt}\right) = \sqrt{\pi Dt}\left(1 + \text{Erf}(R)\right), \tag{3.171}$$

where $\text{Erf}(R)$ is the error function [10] and

$$R = \left(rt - \ln\left(1 - (x_0/K)^\nu\right)\right) / \left(2\sqrt{Dt}\right). \tag{3.172}$$

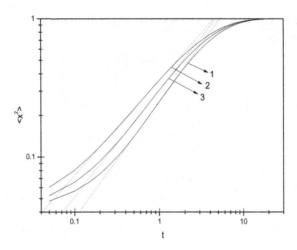

Fig. 3.8    Plots of the second moment $\langle x^2(t) \rangle$ versus time, Eqs. (3.164), (3.165) and (3.170), for the following parameter values: $r = 1$, $x_0 = 0.2$, $K = 1$ and $D = 1$. The solid lines are obtained from Eqs. (3.164), (3.165) and (3.170), whereas the dotted lines correspond to the fitted functions. The curves 1 correspond to the Verhulst logistic model ($\mu = \nu = 1$) and the fitted function $f_1(t) = 0.25t^{0.84}$, the curves 2 correspond to the Gompertz model and the fitted function $f_2(t) = 0.3585t^{0.687}$, and the curves 3 correspond to the generalized logistic model ($\mu = \nu = 0.5$) and the fitted function $f_3(t) = 0.304t^{0.77}$.

The PDF becomes

$$\rho(x,t) = \nu \frac{\exp\left(-\frac{\left(\ln\left(\frac{K^\nu - x^\nu}{\nu K^\nu}\right) - \ln\left(\frac{K^\nu - x_0^\nu}{\nu K^\nu}\right) + rt\right)^2}{4Dt}\right)}{\sqrt{\pi Dt} x^{1-\nu}(K^\nu - x^\nu)(1 + \text{Erf}(R))}. \tag{3.173}$$

The PDF (3.173) also reduces to the Gompertz model for $\nu \to 0$. Note that the consistency of the normalized PDF (3.173) given by $\int_0^K dx \rho(x,t) = 1$ can be verified by numerical calculation for different values of $\nu$ and $t$.

Usually discussions and investigations on the question of interpretation in the Langevin equation with multiplicative white noise are limited to the Ito and Stratonovich prescriptions. In this section, the transformation of the Fokker-Planck equation, which corresponds to the Langevin equation with multiplicative white noise, into the Wiener process has been provided for any prescription. This does not mean that we can obtain solutions for generic drift and diffusion coefficients, and for any prescription. In order to obtain the solution for a given drift and diffusion coefficient, the relation (3.140) must be satisfied.

## 3.5   Similarity solution

Exact solutions for the Fokker-Planck equation are not easy to be obtained, except for particular cases of the drift and diffusion coefficients. There are various methods for solving the Fokker-Planck equation [5] such as transformation of variables, variational approach, reduction to an Hermitian problem, numerical integration, expansion into complete sets, matrix continued-fraction approach, WKB approach and Green's function. Another usefull method concerns the similarity solution [8, 27–29] which reflects scaling and invariant properties of a partial differential equation. Let us first consider the simplest case of the Fokker-Planck equation in one-dimensional space with the vanishing drift coefficient and constant diffusion coefficient,

$$\frac{\partial \rho(x,t)}{\partial t} = D \frac{\partial^2 \rho(x,t)}{\partial x^2}. \tag{3.174}$$

Eq. (3.174) is also called the diffusion equation. This last equation is invariant under the following transformation:

$$x \to \epsilon^a x \text{ and } t \to \epsilon^{2a} t; \tag{3.175}$$

the scalings (3.175) also reflect in the solution of the equation (3.174),

$$\rho(x,t) = \frac{1}{\sqrt{4\pi Dt}} e^{-\frac{1}{4D}\left(\frac{x}{\sqrt{t}}\right)^2}, \tag{3.176}$$

which is scaled as $\rho(x,t) \to \epsilon^{-a}\rho(x,t)$. Thus, the diffusion equation (3.174) has a symmetry transformation $(x,t) \to (\epsilon^a x, \epsilon^{2a} t)$, then a solution of it of the form $\rho = f(z)$ is called a similarity solution and $z = x/\sqrt{t}$ is the similarity variable.

### 3.5.1   *Similarity solution*

Now we consider the general drift and diffusion coefficients. In this case, we assume the following extended scale transformation:

$$\bar{x} = \epsilon^a x \ , \ \ \bar{t} = \epsilon^b t,$$

$$\bar{D}_1(\bar{x}, \bar{t}) = \epsilon^c D_1(x,t) \ , \ \bar{D}_2(\bar{x}, \bar{t}) = \epsilon^d D_2(x,t) \text{ and } \bar{\rho}(\bar{x}, \bar{t}) = \epsilon^e \rho(x,t), \tag{3.177}$$

which includes the drift and diffusion coefficients. Eq. (3.1) is invariant under the scale transformation (3.177) for $b = a - c = 2a - d$. Then, we can seek a similarity solution for Eq. (3.1) with the similarity variable

$$z = \frac{x}{t^\eta}, \tag{3.178}$$

where $\eta = a/b$ and $a, b \neq 0$. Now consider that the PDF is described by

$$\rho(x, t) = t^r \xi(z), \tag{3.179}$$

where $r$ is a real parameter and $\xi(z)$ is a scale-invariant function. Using the scale transformation (3.177) we have

$$\bar{\rho}(\bar{x}, \bar{t}) = \epsilon^e \rho(x, t) = \epsilon^{e-br} \bar{\rho}(\bar{x}, \bar{t}), \tag{3.180}$$

which implies that

$$e = br. \tag{3.181}$$

Substituting Eq. (3.181) into Eq. (3.179) yields

$$\rho(x, t) = t^{\frac{e}{b}} \xi(z) = t^{\frac{\eta e}{a}} \xi(z). \tag{3.182}$$

Further, imposing the normalization of the PDF we can reduce the number of parameters, i.e,

$$\int_{\text{domain}} \rho(x, t) dx = \int_{\text{domain}} t^{\frac{\eta e}{a}} \xi(z) dx = t^{\eta(1+\frac{e}{a})} \int_{\text{domain}} \xi(z) dz = 1, \tag{3.183}$$

which implies that $1 + e/a = 0$, and thus

$$\rho(x, t) = t^{-\eta} \xi(z). \tag{3.184}$$

The above procedure can also be used to determine the drift and diffusion coefficients, and they are given by

$$D_1(x, t) = t^{\eta-1} d_1(z) \quad \text{and} \quad D_2(x, t) = t^{2\eta-1} d_2(z). \tag{3.185}$$

Substituting Eqs. (3.184) and (3.185) into Eq. (3.1) we obtain

$$\frac{d^2}{dz^2} [d_2(z)\xi(z)] + \frac{d}{dz} [\eta z \xi(z) - d_1(z)\xi(z)] = 0. \tag{3.186}$$

Thus, we have used the similarity method to transform the partial differential equation into the ordinary differential equation. Integrating Eq. (3.186) yields

$$\frac{d}{dz} [d_2(z)\xi(z)] + [\eta z - d_1(z)] \xi(z) = C, \tag{3.187}$$

where $C$ is an integration constant. The constant $C$ can be determined by the boundary conditions of the probability density and probability current $S(x, t)$. Eq. (3.187) can be written as follows:

$$S(x, t) = \frac{1}{t} [\eta z \xi(z) - C] = \frac{1}{t} [\eta x \rho(x, t) - C]. \tag{3.188}$$

For natural boundary conditions, the PDF and the probability current must vanish which imply that $C = 0$. With $C = 0$, the solution of Eq. (3.187) is given by

$$\xi(z) = \frac{C_1}{d_2(z)} e^{-\int^z \frac{\eta \bar{z} - d_1(\bar{z})}{d_2(\bar{z})} d\bar{z}}, \qquad (3.189)$$

where $C_1$ is an integration constant. Substituting (3.189) into (3.184) we obtain

$$\rho(x, t) = \frac{C_1}{t^\eta d_2(\frac{x}{t^\eta})} e^{-\int^{x/t^\eta} \frac{\eta \bar{z} - d_1(\bar{z})}{d_2(\bar{z})} d\bar{z}}, \qquad (3.190)$$

where $C_1$ is obtained from the normalization.

As an example we consider the following Fokker-Planck equation:

$$\frac{\partial \rho(x, t)}{\partial t} = D \frac{\partial^2 |x|^{-\theta} \rho(x, t)}{\partial x^2}, \quad \text{for } \theta \neq -2. \qquad (3.191)$$

The interest in this equation is due to the fact that it can describe anomalous diffusion processes. Note that Eq. (3.191) is symmetric under the change $x \to -x$, and it corresponds to the Ito prescription with the following drift and diffusion coefficients:

$$D_1(x, t) = h_1(x, t) = 0 \quad \text{and} \quad D_2(x, t) = h_2^2(x, t) = D|x|^{-\theta}. \qquad (3.192)$$

Besides, the diffusion coefficient (3.192) has been applied to several physical situations such as fast electrons in a hot plasma in the presence of dc electric fields, turbulent two-particle diffusion in configuration space; also, both the Richardson and the Kolmogorov laws applied it to turbulence and to describe the diffusion on fractals. From Eq. (3.185), we obtain the functions $d_1(z)$ and $d_2(z)$, and they are given by

$$d_1(z) = 0 \quad \text{and} \quad d_2(z) = D|z|^{-\theta}, \qquad (3.193)$$

with $\eta = 1/(2 + \theta)$. The solution for the PDF is obtained by substituting (3.193) into (3.190), and we have

$$\rho(x, t) = \frac{C_1 |x|^\theta}{D t^{\frac{1+\theta}{2+\theta}}} e^{-\frac{|x|^{2+\theta}}{D(2+\theta)^2 t}}. \qquad (3.194)$$

Now we want to find a solution for the Fokker-Planck equation (3.191) with the coefficient $h_2(x, t)$ given by (3.192) in any prescription; the drift and diffusion coefficients are given by

$$D_1(x, t) = 2\lambda h_2(x, t) \frac{\partial h_2(x, t)}{\partial x} \quad \text{and} \quad D_2(x, t) = h_2^2(x, t) = D|x|^{-\theta}. \qquad (3.195)$$

The Fokker-Planck equation with the coefficients given by (3.195) is also symmetric under the change $x \to -x$. Thus, we can solve the equation for the interval $(0, \infty)$, and the solution for the whole space is formed from the solutions of two intervals: $(-\infty, 0)$ and $(0, \infty)$. For $x > 0$ we have

$$d_1(z) = -\lambda D\theta z^{-(1+\theta)} \quad \text{and} \quad d_2(z) = Dz^{-\theta}, \qquad (3.196)$$

with $\eta = 1/(2 + \theta)$. The solution for the PDF is given by

$$\rho(x,t) = \frac{|x|^{(1-\lambda)\theta}}{2|2+\theta|^{\frac{(1-2\lambda)\theta}{2+\theta}} \Gamma\left(\frac{1+(1-\lambda)\theta}{2+\theta}\right) (Dt)^{\frac{1+(1-\lambda)\theta}{2+\theta}}} e^{-\frac{|x|^{2+\theta}}{D(2+\theta)^2 t}}, \qquad (3.197)$$

with

$$\frac{1+(1-\lambda)\theta}{2+\theta} > 0. \qquad (3.198)$$

The PDF (3.197) describes the stretched ($-2 < \theta < 0$) and compressed ($\theta > 0$) Gaussian distribution only for the Ito description ($\lambda = 1$). For $\lambda \neq 1$ the PDF is the Weibull distribution, and the power-law $|x|^{(1-\lambda)\theta}$ may split the peak of the compressed ($\theta > 0$) Gaussian distribution into two ones; for $\theta < 0$ the PDF diverges at $x = 0$ (see Fig. 3.9). Note that the prescription parameter $\lambda$ appears only on the power-law. Thus, the exponential decay of the distribution for a fixed time is mainly controlled by the parameters $D$ and $\theta$.

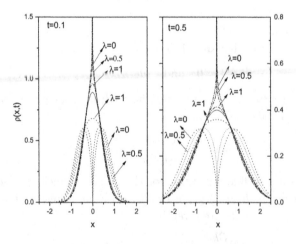

Fig. 3.9 Plots of the PDF (3.197) for $D = 1$. The dash-dotted lines correspond to $\theta = -0.1$, the solid lines correspond to $\theta = 0$, and the dotted lines correspond to $\theta = 0.5$.

The $n$-moment (even $n$) related to the PDF (3.197) is given by

$$\langle x^n \rangle = \frac{\left[ D \left( 2 + \theta \right)^2 \right]^{\frac{n}{2+\theta}} \Gamma \left( \frac{1+n+(1-\lambda)\theta}{2+\theta} \right) t^{\frac{n}{2+\theta}}}{\Gamma \left( \frac{1+(1-\lambda)\theta}{2+\theta} \right)}. \tag{3.199}$$

Note that the second moment given by Eq. (3.199) can describe subdiffusive, normal and superdiffusive processes. Besides, the prescription parameter $\lambda$ can change the behavior of the PDF (3.197), but it does not change the behavior of the $n$-moment (3.199) due to the fact that $t^{\frac{n}{2+\theta}}$ does not depend on $\lambda$.

## 3.6 Solution in a finite interval and first passage time

When a system is constrained in a finite interval, it may be subject to different kinds of boundary conditions. It is often of interest to know how long a particle remains in a certain region of space; a physical quantity which may be related to this time is the mean first passage time (MFPT). In particular, we consider a particle diffusing in a finite interval $[a, b]$, whose dynamics is described by the Fokker-Planck equation (3.1), subjects to the absorbing boundaries

$$\rho(a, t) = \rho(b, t) = 0 \tag{3.200}$$

and the initial condition

$$\rho(x, 0) = \delta(x - x_0). \tag{3.201}$$

### 3.6.1 *First passage time distribution and mean first passage time*

First passage time (FPT) is the time at which a process, starting from a given point, leaves a given domain $[a, b]$ for the first time. In terms of the Fokker-Planck equation we want to obtain the probability density $\rho(x, t)$ which is for the stochastic variable $\xi(t)$ starting at $t = 0$ with $\xi(0) = x_0$ to reach $x$ at time $t$. If we consider that the particles are no longer counted when they have passed a boundary, then the boundary can be replaced by an absorbing boundary with $P(x, t | x_0, t_0)|_{boundary} = 0$. In this case, we impose the following conditions for the Fokker-Planck equation:

$$\frac{\partial P}{\partial t} = L_{FP}(x, t) P; \quad P(x, 0 | x_0, 0) = \delta(x - x_0) \quad \text{for} \quad a < x < b, \tag{3.202}$$

$$P(x, t | x_0, 0) = 0 \quad \text{for} \quad x = a \quad \text{or} \quad x = b. \tag{3.203}$$

The FPT distribution in terms of the probability density $P(x, t|x_0, 0)$ [5] is calculated by

$$\mathcal{F}(t) = -\frac{d}{dt} \int_a^b P(x, t|x_0, 0)dx, \qquad (3.204)$$

whereas the mean first passage time [30] is calculated by

$$M = \int_0^\infty \int_a^b P(x, t|x_0, 0)dxdt. \qquad (3.205)$$

### 3.6.2  *Solution for vanishing drift coefficient and constant diffusion coefficient*

For vanishing drift coefficient and constant diffusion coefficient

$$D_1(x, t) = 0 \quad \text{and} \quad D_2(x, t) = D, \qquad (3.206)$$

an exact solution for $\rho(x, t)$ can be obtained from Eq. (3.1) by using the method of separation of variables [5, 31–34], i.e,

$$\rho(x, t) = \rho_x(x)\rho_t(t) . \qquad (3.207)$$

Substituting Eq. (3.207) into Eq. (3.1) yields

$$\frac{d\rho_t(t)}{dt} = -\mu\rho_t(t), \qquad (3.208)$$

and

$$\frac{d^2\rho_x(x)}{dx^2} = -\frac{\mu}{D}\rho_x(x) \qquad (3.209)$$

where $\mu$ is the separation constant. The solution of Eq. (3.208) is given by

$$\rho_t(t) = e^{-\mu t}, \qquad (3.210)$$

whereas the solution of Eq. (3.209) may be expressed by

$$\rho_x(x) = A\sin(Bx + C), \qquad (3.211)$$

with

$$B^2 = \frac{\mu}{D}. \qquad (3.212)$$

Applying the boundary conditions (3.200) yields

$$\rho_x(a) = A\sin(Ba + C) = 0 \quad \text{and} \quad \rho_x(b) = A\sin(Bb + C) = 0. \qquad (3.213)$$

Note that Eq. (3.213) is satisfied for $Ba + C = Bb + C = n\pi$ and $n = 1, 2, ...$, but the relation gives a trivial solution $a = b$. For non-trivial solution we choose

$$Ba + C = 0 \quad \text{and} \quad Bb + C = n\pi, \quad n = 1, 2, ..., \qquad (3.214)$$

thus we have

$$B = \frac{n\pi}{b-a} \quad \text{and} \quad C = -\frac{n\pi a}{b-a}. \tag{3.215}$$

Substituting Eq. (3.215) into Eq. (3.211) we obtain

$$\rho_{xn}(x) = A_n \sin\left(\frac{n\pi(x-a)}{b-a}\right), \tag{3.216}$$

with

$$\mu_n = D\left(\frac{n\pi}{b-a}\right)^2. \tag{3.217}$$

The solution for $\rho(x,t)$ is given by a sum of the solutions for space and time parts, i.e,

$$\rho(x,t) = \sum_{n=1}^{\infty} A_n \sin\left(\frac{n\pi(x-a)}{b-a}\right) e^{-\mu_n t}. \tag{3.218}$$

The coefficients $A_n$ are determined by imposing the initial condition $\rho(x,0) = \delta(x - x_0)$ as follows:

$$\int_a^b \delta(x-x_0) \sin\left(\frac{k\pi(x-a)}{b-a}\right) dx =$$

$$\sum_{n=1}^{\infty} A_n \int_a^b \sin\left(\frac{k\pi(x-a)}{b-a}\right) \sin\left(\frac{n\pi(x-a)}{b-a}\right) dx, \tag{3.219}$$

where $k$ is a positive integer number. Integrating Eq. (3.219) yields

$$A_n = \frac{2}{b-a} \sin\left(\frac{n\pi(x_0-a)}{b-a}\right). \tag{3.220}$$

Substituting Eq. (3.220) into Eq. (3.218) we arrive at

$$\rho(x,t) = \frac{2}{b-a} \sum_{n=1}^{\infty} \sin\left(\frac{n\pi(x_0-a)}{b-a}\right) \sin\left(\frac{n\pi(x-a)}{b-a}\right) e^{-D\left(\frac{n\pi}{b-a}\right)^2 t}. \tag{3.221}$$

For the interval $[0, L]$ and $x_0 = L/2$, the first passage time distribution has the following result:

$$\mathcal{F}(t) = \frac{4D\pi}{L^2} \sum_{n=0}^{\infty} (-1)^n (1+2n) e^{-D\left(\frac{\pi(1+2n)}{L}\right)^2 t}, \tag{3.222}$$

whereas the mean first passage time reduces to a simple expression given by

$$M = \frac{L^2}{8D}. \tag{3.223}$$

### 3.6.3    *Solution for power-law diffusion coefficient*

We now consider the coefficients $D_1(x,t)$ and $D_2(x,t)$ given by (3.195) in any prescription and we will find a solution for the Fokker-Planck equation (3.1) in the interval $[0, L]$. Applying the separation of variables

$$\rho(x,t) = \rho_t(t)\rho_x(x) \tag{3.224}$$

to the Fokker-Planck equation with the coefficients (3.195) we obtain

$$\rho_t(t) = e^{-\mu_n Dt} \tag{3.225}$$

and

$$\frac{d}{dx}\left[|x|^{-\lambda\theta}\frac{d}{dx}\left(|x|^{-(1-\lambda)\theta}\rho_x(x)\right)\right] = -\mu_n\rho_x(x), \tag{3.226}$$

where $\mu_n$ is the separation constant. Eq. (3.226) can be written as the following form:

$$\frac{d^2\rho_x(x)}{dx^2} - (2-\lambda)\theta x^{-1}\frac{d\rho_x(x)}{dx} + \left[(1-\lambda)\theta(1+\theta) + \mu_n x^{2+\theta}\right]x^{-2}\rho_x(x) = 0. \tag{3.227}$$

A solution of this last equation may be given in terms of the Bessel function [6]

$$\rho_x(x) = x^{\frac{1+(2-\lambda)\theta}{2}}J_\nu\left(2\sqrt{\frac{\mu_n}{(2+\theta)^2}}x^{\frac{2+\theta}{2}}\right), \tag{3.228}$$

where $J_\nu(z)$ is the Bessel function of the first kind and $\nu$ is given by

$$\nu = \frac{\sqrt{(1+(2-\lambda)\theta)^2 - 4(1-\lambda)\theta(1+\theta)}}{2+\theta}. \tag{3.229}$$

In order to satisfy the boundary conditions the solution $\rho(x,t)$ should be given in terms of the expansion of the eigenfunctions

$$\rho(x,t) = \sum_{n=1}^{\infty} C_n\rho_t(t)\rho_x(x)$$

$$= \sum_{n=1}^{\infty} C_n x^{\frac{1+(2-\lambda)\theta}{2}}J_\nu\left(2\sqrt{\frac{\mu_n}{(2+\theta)^2}}x^{\frac{2+\theta}{2}}\right)e^{-\mu_n Dt}, \tag{3.230}$$

where $C_n$ are the coefficients of the expansion determined by imposing the initial condition $\rho(x,0) = \delta(x-x_0)$ as follows:

$$\int_0^L x^A\delta(x-x_0)J_\nu\left(2\sqrt{\frac{\mu_k}{(2+\theta)^2}}x^{\frac{2+\theta}{2}}\right)dx$$

$$= \sum_{n=1}^{\infty} C_n \int_0^L x^{A + \frac{1+(2-\lambda)\theta}{2}} J_\nu \left( 2\sqrt{\frac{\mu_k}{(2+\theta)^2}} x^{\frac{2+\theta}{2}} \right) J_\nu \left( 2\sqrt{\frac{\mu_n}{(2+\theta)^2}} x^{\frac{2+\theta}{2}} \right) dx,$$

$$(3.231)$$

where $A$ is a constant. Introducing a new variable

$$u = \left( \frac{x}{L} \right)^{\frac{2+\theta}{2}},$$

$$(3.232)$$

we obtain

$$x_0^A J_\nu \left( 2\sqrt{\frac{\mu_k}{(2+\theta)^2}} x_0^{\frac{2+\theta}{2}} \right) = \frac{2L^{\frac{2A+3+(2-\lambda)\theta}{2}}}{2+\theta}$$

$$\times \sum_{n=1}^{\infty} C_n \int_0^1 u^{\frac{1+2A+(1-\lambda)\theta}{2+\theta}} J_\nu \left( 2\sqrt{\frac{\mu_k}{(2+\theta)^2}} L^{\frac{2+\theta}{2}} u \right) J_\nu \left( 2\sqrt{\frac{\mu_n}{(2+\theta)^2}} L^{\frac{2+\theta}{2}} u \right) du.$$

$$(3.233)$$

Using the relation [6]

$$\int_0^1 u J_\nu (\alpha u) J_\nu (\beta u) \, du$$

$$= \begin{cases} 0, & (\alpha \neq \beta) \\ \frac{(J_{1+\nu}(\alpha))^2}{2}; & (\alpha = \beta); [J_\nu (\alpha) = J_\nu (\beta) = 0, \nu > -1] \end{cases}$$

$$(3.234)$$

and setting

$$\frac{1 + 2A + (1-\lambda)\theta}{2+\theta} = 1$$

$$(3.235)$$

we obtain

$$C_n = \frac{(2+\theta) x_0^{\frac{1+\lambda\theta}{2}} J_\nu \left( \omega_n \left( \frac{x_0}{L} \right)^{\frac{2+\theta}{2}} \right)}{L^{2+\theta} J_{1+\nu}^2 (\omega_n)},$$

$$(3.236)$$

and

$$\omega_n = 2\sqrt{\frac{\mu_n}{(2+\theta)^2}} L^{\frac{2+\theta}{2}}$$

$$(3.237)$$

are associated with the zeros of the Bessel function $J_\nu (\omega_n) = 0$. Substituting Eq. (3.236) into Eq. (3.230) we arrive at

$$\rho(x,t) = \frac{(2+\theta) x_0^{\frac{1+\lambda\theta}{2}}}{L^{2+\theta}}$$

$$\times \sum_{n=1}^{\infty} \frac{J_\nu \left( \omega_n \left( \frac{x_0}{L} \right)^{\frac{2+\theta}{2}} \right) x^{\frac{1+(2-\lambda)\theta}{2}} J_\nu \left( \omega_n \left( \frac{x}{L} \right)^{\frac{2+\theta}{2}} \right) e^{-\frac{D(2+\theta)^2 \omega_n^2 t}{4L^{2+\theta}}}}{J_{1+\nu}^2 (\omega_n)}.$$

$$(3.238)$$

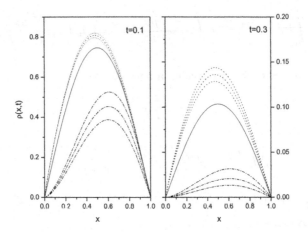

Fig. 3.10   Plots of the PDF (3.238) for $t = 0.1, 0.3$. $\theta = -0.1, 0, 0.5$, $\lambda = 0, 0.5, 1$, $D = 1$, $L = 1$ and $x_0 = L/2$. The dotted lines correspond to $\theta = -0.1$ and $\lambda = 0, 0.5, 1$ from bottom to top. The solid lines correspond to $\theta = 0$, and the dash-dotted lines correspond to $\theta = 0.5$ and $\lambda = 0, 0.5, 1$ from top to bottom

Fig. 3.10 shows the PDF (3.238) in the finite interval $L = 1$. The PDF presents asymmetry for $\theta \neq 0$. Moreover, the peak of the PDF increases with the increase of $\lambda$ for $\theta = -0.1$, and it decreases with the increase of $\lambda$ for $\theta = 0.5$.

The first passage time distribution and mean first passage time, related to the PDF (3.238), are given by

$$\mathcal{F}(t) = \frac{D(2+\theta)^2}{2L^{2+\theta}} \left(\frac{x_0}{L}\right)^{\frac{1+\lambda\theta}{2}} \sum_{n=1}^{\infty} \frac{\omega_n^2 J_\nu \left(\omega_n \left(\frac{x_0}{L}\right)^{\frac{2+\theta}{2}}\right)}{J_{1+\nu}^2(\omega_n)}$$

$$\times \left[\frac{2^{\alpha_0}\Gamma\left(\frac{1+\alpha_0+\nu}{2}\right)}{\omega_n^{1+\alpha_0}\Gamma\left(\frac{1-\alpha_0+\nu}{2}\right)} - \frac{J_{\nu-1}(\omega_n) S_{\alpha_0,\nu}(\omega_n)}{\omega_n^{\alpha_0}}\right] e^{-\frac{D(2+\theta)^2\omega_n^2 t}{4L^{2+\theta}}} \quad (3.239)$$

and

$$M = \frac{8L^{2+\theta}}{D(2+\theta)^2} \left(\frac{x_0}{L}\right)^{\frac{1+\lambda\theta}{2}} \sum_{n=1}^{\infty} \frac{J_\nu \left(\omega_n \left(\frac{x_0}{L}\right)^{\frac{2+\theta}{2}}\right)}{\omega_n^2 J_{1+\nu}^2(\omega_n)}$$

$$\times \left[\frac{2^{\alpha_0}\Gamma\left(\frac{1+\alpha_0+\nu}{2}\right)}{\omega_n^{1+\alpha_0}\Gamma\left(\frac{1-\alpha_0+\nu}{2}\right)} - \frac{J_{\nu-1}(\omega_n) S_{\alpha_0,\nu}(\omega_n)}{\omega_n^{\alpha_0}}\right], \quad (3.240)$$

where $S_{\mu,\nu}(z)$ is the Lommel function [6] and

$$\alpha_0 = \frac{1 + (1 - \lambda)\theta}{2}. \qquad (3.241)$$

Fig. 3.11 shows the FPT distribution (3.239) in the finite interval $L = 2$ for different values of $\lambda$ and $\theta$; it exhibits different behaviors for different values of $\lambda$.

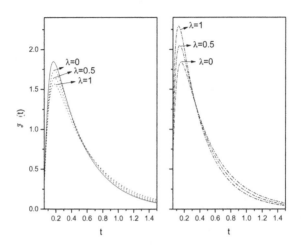

Fig. 3.11   Plots of the FPT distribution (3.239) for $D = 1$, $L = 2$ and $x_0 = L/2$. The dotted lines correspond to $\theta = -0.3$, the solid line corresponds to $\theta = 0$, and the dash-dotted lines correspond to $\theta = 0.5$.

## 3.7 Langevin equation with multiplicative noise in different orders of prescription and its connection with the Tsallis distribution

We now discuss on the Fokker-Planck equation and its connection between the stationary solution and the Tsallis distribution for different orders of prescription in discretization rules for the stochastic integrals; the Tsallis index $q$ and the prescription parameter $\lambda$ are determined with the drift and diffusion coefficients. The result is quite general. For application, one shows that the Tsallis distribution can be described by a class of population growth models subject to the linear multiplicative white noise.

The Tsallis distribution [35, 36] has been connected to a variety of natural systems (see, for instance, [37–42]), and it is given by

$$\rho_q(x) = N \left[ 1 - \beta(1 - q)U(x) \right]^{\frac{1}{1-q}}, \qquad (3.242)$$

where $U(x)$ is the potential. The distribution (3.242) is obtained by extremizing the Tsallis's entropy

$$S_q = k\frac{1 - \sum_{i=1}^{w} p_i^q}{q - 1} \tag{3.243}$$

with the constraints

$$E_q = \sum_{i=1}^{w} p_i^q U_i \tag{3.244}$$

and

$$\sum_{i=1}^{w} p_i = 1, \tag{3.245}$$

where $k$ is a positive constant, $w$ is the total number of microscopic configurations of the system, $q$ is a real parameter, $p_i$ is the probability and $E_q$ is the generalized internal energy. Notice that this result extends the one given in Ref. [43]. Moreover, the Tsallis distribution can be described by a class of population growth models subject to the linear multiplicative white noise.

In one-dimensional space, the forward Fokker-Planck equation [1–3,5,12] for the probability distribution corresponding to the Langevin equation (2.1) with the deterministic drift and multiplicative noise term independent of the time is given by

$$\frac{\partial \rho(x,t)}{\partial t} = -\frac{\partial}{\partial x}\left[D_1(x)\rho(x,t)\right] + \frac{\partial^2}{\partial x^2}\left[D_2(x)\rho(x,t)\right] , \tag{3.246}$$

where $D_1(x)$ and $D_2(x)$ are the drift and diffusion coefficients given by

$$D_1(x) = h_1(x) + 2\lambda\frac{dh_2(x)}{dx}h_2(x) \tag{3.247}$$

and

$$D_2(x) = h_2^2(x), \tag{3.248}$$

and $\lambda$ ($0 \leq \lambda \leq 1$) is the prescription parameter due to the discretization rules for the stochastic integrals. The stationary solution of Eq. (3.246) can be obtained, and it is given by

$$\rho_{st}(x) = N\exp\left(\int \frac{dx}{D_2(x)}\left(D_1(x) - \frac{dD_2(x)}{dx}\right)\right) . \tag{3.249}$$

This last distribution is equal to the Tsallis distribution (3.242) if the following relation is satisfied:

$$\frac{1}{D_2(x)}\left[h_1(x) + (\lambda - 1)\frac{dD_2(x)}{dx}\right] = -\frac{\beta}{1 - \beta(1-q)U(x)}\frac{dU(x)}{dx} . \tag{3.250}$$

Note that this last relation is quite general, i.e, it is valid for a large class of drift and diffusion coefficients and any prescription.

### 3.7.1 Tsallis distribution and the atom-laser interaction in the optical lattice

Now we show that the probability distribution in momentum space of the atom-laser interaction in the optical lattice is described by the Tsallis distribution [44]. The Rayleigh equation that corresponds to the semiclassical Wigner function $\rho(p, t)$ can be obtained from the microscopic Hamiltonian that describes the atom-laser interaction in the optical lattice [45], and it is given by

$$\frac{\partial \rho(p, t)}{\partial t} = -\frac{\partial}{\partial p}\left[K(p)\rho(p, t)\right] + \frac{\partial}{\partial p}\left[D(p)\frac{\partial \rho(p, t)}{\partial p}\right], \qquad (3.251)$$

where $p$ is the momentum, $K(p)$ and $D(p)$ are the drift and diffusion coefficients given by

$$K(p) = -\frac{\alpha p}{1 + \left(\frac{p}{p_c}\right)^2} \qquad (3.252)$$

and

$$D(p) = D_0 + \frac{D_1}{1 + \left(\frac{p}{p_c}\right)^2}. \qquad (3.253)$$

The drift coefficient $K(p)$ represents a cooling force (due to the Sisyphus effect) with a damping coefficient $\alpha$. This force acts only on slow particles with a momentum smaller than the capture momentum $p_c$. The diffusion coefficient $D(p)$ describes stochastic momentum fluctuations and accounts for heating processes. Eq. (3.251) is valid for (1) low intensity of the laser, (2) high velocity of the atoms $mv^2/2 \gg U_0$, where $U_0$ is the potential depth, and (3) $p \gg \hbar k$, where $k$ is the wave number of the laser field.

Eqs. (3.252) and (3.253) can be written as follows:

$$\frac{K(p)}{D(p)} = -\frac{\alpha^*}{1 - \alpha^*(1 - q)U(p)}\frac{dU(p)}{dp}, \qquad (3.254)$$

where

$$\alpha^* = \frac{\alpha}{2\left(D_0 + D_1\right)}, \quad q = 1 + \frac{2D_0}{\alpha p_c^2} \quad \text{and} \quad U(p) = p^2. \qquad (3.255)$$

Eq. (3.254) is equal to Eq. (3.250) only for $\lambda = 1$. Thus, Eq. (3.251) with the drift and diffusion coefficients given by Eqs. (3.252) and (3.253) can be described by the Langevin equation (2.1) with the transport prescription .

### 3.7.2   *Tsallis distribution and a class of population growth models with linearly coupled noise*

For another application of Eq. (3.250), we consider the models of population growth described in Sec. 3.4.3

$$h_1(x) = r \frac{x \left[1 - \left(\frac{x}{K}\right)^\nu\right]}{\mu \left[1 - \left(1 - \frac{\nu}{\mu}\right)\left(\frac{x}{K}\right)^\nu\right]} , \qquad (3.256)$$

where $x(t)$ is the number of population alive at time $t$, $\mu$ and $\nu$ are positive real parameters, $r$ is the intrinsic growth rate and $K$ is the carrying capacity. The deterministic model given by Eqs. (2.1) and (3.256), with $h_2(x) = 0$, contains the classical growth models. In applications, the influence of the noise (such as a change in temperature, food and water supplies) is usually considered by the linear multiplicative noise of the effective birth rate [46] which is given by

$$h_2(x) = \epsilon x, \qquad (3.257)$$

where $\epsilon$ is associated with the noise strength; in this case, the diffusion coefficient is given by $D_2(x) = \epsilon^2 x^2$. For nonlinearly coupled noise see for instance [47, 48].

Now, we consider the linearly coupled noise; from Eq. (3.249) yields

$$\rho_{st}(x) \sim x^{\frac{r}{\mu\epsilon^2} + 2(\lambda - 1)} \left[1 - \left(1 - \frac{\nu}{\mu}\right)\left(\frac{x}{K}\right)^\nu\right]^{\overline{\mu\epsilon^2(\mu-\nu)}} . \qquad (3.258)$$

In particular, for $\mu = \nu$ one obtains

$$\rho_{st}(x) \sim x^{\frac{r}{\mu\epsilon^2} + 2(\lambda - 1)} \exp\left(-\frac{r}{\mu^2\epsilon^2}\left(\frac{x}{K}\right)^\mu\right) . \qquad (3.259)$$

The distribution (3.259) is well-known and it is called the Weibull distribution. Besides, the distribution (3.258) presents interesting aspects. In order to retain a consistent probabilistic interpretation, the cut-off condition imposes $\rho(x) = 0$ whenever $\mu > \nu$ and $\left[1 - \left(1 - \frac{\nu}{\mu}\right)\left(\frac{x}{K}\right)^\nu\right] < 0$. For $\lambda = 1$ and positive values for $r$ and $\mu$ the distribution (3.258) is zero at $x = 0$; in this case the populations can always survive. However, for $\lambda \neq 1$ the distribution (3.258) is divergent at $x = 0$ for

$$-1 < \frac{r}{\mu\epsilon^2} + 2(\lambda - 1) < 0 ; \qquad (3.260)$$

this means that the order of prescription in the discretization rules for the stochastic integrals may prevent population extinction.

The Tsallis distribution can be obtained from the solution (3.258), by setting

$$\frac{r}{\mu\epsilon^2} + 2\left(\lambda - 1\right) = 0, \quad \lambda \neq 1 \tag{3.261}$$

and

$$1 - q = \frac{\mu\epsilon^2\left(\mu - \nu\right)}{r}. \tag{3.262}$$

Eq. (3.261) implies that

$$0 < \frac{r}{2\mu\epsilon^2} \leq 1 \tag{3.263}$$

and

$$\lambda = 1 - \frac{r}{2\mu\epsilon^2}. \tag{3.264}$$

The expression (3.262) shows that the Tsallis index $q$ can assume any real value, and it is also associated with the microscopic parameters of the system. For $q = 1$ yields $\mu = \nu$, and the distribution (3.258) reduces to the stretched exponential for $0 < \mu < 1$, exponential function for $\mu = 1$ and compressed exponential for $\mu > 1$. It is worth noting that the distribution (3.258) with the conditions (3.261) and (3.262) also satisfies the relation (3.250) for

$$U(x) = x^\nu \tag{3.265}$$

and

$$\beta = \frac{r}{\mu^2\epsilon^2 K^\nu}. \tag{3.266}$$

In summary, the stationary solution of the Fokker-Planck equation has been connected with the Tsallis distribution for generic drift and diffusion coefficients. The stationary solution of a class of population growth models subject to the linear multiplicative white noise could be described by the Tsallis distribution for different orders of prescription. In these models, the Tsallis index $q$ has been connected with the microscopic parameters. Interesting aspects have been achieved; the orders of prescription in discretization rules for the stochastic integrals play a key role to connect with the Tsallis distribution. Moreover, the change in the prescription parameter $\lambda$ of a given system may lead to the extinction or survival of a population. As is expected, the prescription parameter can modify the behavior of the system considerably, i.e, different prescription may describe different behavior.

# Bibliography

[1]  P. Hänggi and H. Thomas, *Phys. Rep.* **88**, 207 (1982).

[2]  P. Hänggi, *Helv. Phys. Acta* **51**, 183 (1978).

[3]  K. S. Fa, *Phys. Rev. E* **72**, 020101 (2005).

[4]  J. M. Sancho, *Phys. Rev. E* **84**, 062102 (2011).

[5]  H. Risken, *The Fokker-Planck Equation*, second ed. (Springer-Verlag, Berlin, 1996).

[6]  I. S. Gradshteyn and I. M. Ryzhik, *Table of Integrals, Series and Products* (Academic Press, USA, 1965).

[7]  J. Mathews and R. L. Walker, *Mathematical Methods of Physics*, second ed. (Addison-Wesley, USA, 1971).

[8]  R. Haberman, *Applied Partial Differential Equations*, fourth ed. (Pearson, USA, 2004).

[9]  F. Lillo and R. N. Mantegna, *Phys. Rev. E* **61**, R4675 (2000).

[10]  S. Wolfram, *Wolfram Mathematica 9* (Wolfram Media, 2012).

[11]  K. S. Fa, *Chem. Phys.* **287**, 1 (2003).

[12]  K. S. Fa, *Ann. Phys.* **327**, 1989 (2012).

[13]  L. M. Ricciardi, *J. Math Analysis Applic.* **54**, 185 (1976).

[14]  G. Albano, V. Giorno, P. Román-Román and F. Torres-Ruiz, *J. Theo. Bio.* **276**, 67 (2011).

[15]  S. Sakanoue, *Ecol. Modelling* **205**, 159 (2007).

[16]  N. S. Goel, S. C. Maitra and E. W. Montroll, *Rev. Modern Phys.* **43**, 231 (1971).

[17]  E. Renshaw, *Modelling Biological Populations in Space and Time* (Cambridge Univ. Press, Cambridge, 1991).

[18]  R. M. Nisbet and W. S. C. Gurney, *Modelling Fluctuating Populations* (John Wiley & Sons, USA, 1982).

[19]  H. T. Davis, *Introduction to nonlinear differential and integral equations* (Dover, New York, 1962).

[20]  S. I. Rubinow, *Introduction to mathematical biology* (Dover, New York, 2002).

[21]  C. P. D. Birch, *Ann. Botany* **83**, 713 (1999).

[22]  A. Tsoularis and J. Wallace, *Math. Biosci.* **179**, 21 (2002).

[23]  M. Peleg, M. G. Corradini and M. D. Normand, *Food Research Int.* **40**, 808 (2007).

[24]  M. Eigen, *Naturwissenschaften* **58**, 465 (1971).

[25]  P. J. Jackson et al., *Phys. Rev. A* **40**, 2875 (1989).

[26]  G. Aquino, M. Bologna and H. Calisto, *EPL* **89**, 50012 (2010).

[27]  G. W. Bluman and J. D. Cole, *Similarity Methods for Differential Equations* (Springer-Verlag, 1974).

[28]  C. L. Ho, *Ann. Phys.* **327**, 386 (2012).

[29]  C. L. Ho and R. Sasaki, *J. Math. Phys.* **55**, 113301 (2014).

[30]  C. W. Gardiner, *Handbook of Stochastic Methods* (Springer-Verlag, Berlin, 1997).

[31]  R. Metzler, E. Barkai and J. Klafter, *Phys. Rev. Lett.* **82**, 3563 (1999).

[32] K. S. Fa, *J. Chem. Phys.* **137**, 234102 (2012).

[33] G. Rangarajan and M. Ding, *Phys. Lett. A* **273**, 322 (2000).

[34] G. Rangarajan and M. Ding, *Phys. Rev. E* **62**, 120 (2000).

[35] C. Tsallis, *J. Stat. Phys.* **52**, 479 (1988).

[36] C. Tsallis, R. S. Mendes and A. R. Plastino, *Physica A* **261**, 534 (1998).

[37] A. R. Plastino and A. Plastino, *Phys. Lett. A* **174**, 384 (1993).

[38] A. R. Plastino and A. Plastino, *Physica A* **222**, 347 (1995).

[39] M. L. Lyra and C. Tsallis, *Phys. Rev. Lett.* **80**, 53 (1998).

[40] C. Beck, *Phys. Rev. Lett.* **87**, 180601 (2001).

[41] C. Tsallis, M. Gell-Mann and Y. Sato, *PNAS* **102** 15377 (2005).

[42] C. Zander and A. R. Plastino, *Physica A* **364** 145 (2006).

[43] L. Borland, *Phys. Lett. A* **245**, 67 (1998).

[44] E. Lutz, *Phys. Rev. A* **67**, 051402 (2003).

[45] J. Dalibard and C. Cohen-Tannoudji, *J. Opt. Soc. Am. B* **6**, 2023 (1989).

[46] R. Zygadlo, *Phys. Rev. E* **47**, 4067 (1993).

[47] R. Zygadlo, *Phys Rev. E* **54**, 5964 (1996).

[48] H. Calisto and M. Bologna, *Phys. Rev. E* **75**, 050103 (2007).

# Chapter 4

# Fokker-Planck equation for several variables

## 4.1 Introduction

Indubitably, it is usually more difficult to solve the Fokker-Planck equation for several variables than for one variable [1–4]. Some mathematical tools such as matrices and coordinate transformations play important roles and they are largely employed for solving $N$-dimensional differential equations. For time-independent drift and diffusion coefficients the Fokker-Planck equation for $N$ variables is described by

$$\frac{\partial \rho\left(\{\mathbf{x}\},t\right)}{\partial t} = \left[-\frac{\partial}{\partial x_i}D_i\left(\{\mathbf{x}\}\right) + \frac{\partial^2}{\partial x_i x_j}D_{ij}\left(\{\mathbf{x}\}\right)\right]\rho\left(\{\mathbf{x}\},t\right)$$

$$= -\frac{\partial S_i\left(\{\mathbf{x}\},t\right)}{\partial x_i}, \tag{4.1}$$

where $S_i$ is the probability current given by

$$S_i\left(\{\mathbf{x}\},t\right) = \left[D_i\left(\{\mathbf{x}\}\right) - \frac{\partial}{\partial x_j}D_{ij}\left(\{\mathbf{x}\}\right)\right]\rho\left(\{\mathbf{x}\},t\right). \tag{4.2}$$

It should be noted that for $N$ variables the probability current is not necessarily a constant in the stationary state such as for the one-variable case (for more detailed discussions see Ref. [1]). In particular, for the constant diffusion coefficient $D_{ij} = D\delta_{ij}$ the probability current reduces to a simple form

$$S_i\left(\{\mathbf{x}\},t\right) = \rho\left(\{\mathbf{x}\},t\right)\left[D_i\left(\{\mathbf{x}\}\right) - D\frac{\partial \ln\left(\rho\left(\{\mathbf{x}\},t\right)\right)}{\partial x_i}\right]. \tag{4.3}$$

From Eq. (4.3) we see that the stationary state can vanish. By setting $S_i = 0$ we obtain

$$D_i\left(\{\mathbf{x}\}\right) = D\frac{\partial \ln\left(\rho\left(\{\mathbf{x}\}\right)\right)}{\partial x_i}. \tag{4.4}$$

67

The right-hand side term of Eq. (4.4) can be written as a gradient of a potential, then we have

$$D_i\left(\{\mathbf{x}\}\right) = -D\frac{\partial\Phi\left(\{\mathbf{x}\}\right)}{\partial x_i}, \tag{4.5}$$

with

$$\rho_{st}\left(\{\mathbf{x}\}\right) = Ae^{-\Phi(\{\mathbf{x}\})}. \tag{4.6}$$

Integrating Eq. (4.5) yields

$$\Phi\left(\{\mathbf{x}\}\right) = -\frac{1}{D}\int D_i\left(\{\mathbf{x}'\}\right)dx_i'. \tag{4.7}$$

From Eq. (4.5) we obtain the following necessary and sufficient conditions for the existence of the potential $\Phi\left(\{\mathbf{x}\}\right)$:

$$\frac{\partial D_i\left(\{\mathbf{x}\}\right)}{\partial x_j} = \frac{\partial D_j\left(\{\mathbf{x}\}\right)}{\partial x_i}, \tag{4.8}$$

since

$$\frac{\partial^2\Phi\left(\{\mathbf{x}\}\right)}{\partial x_i x_j} = \frac{\partial^2\Phi\left(\{\mathbf{x}\}\right)}{\partial x_j x_i}. \tag{4.9}$$

## 4.2  Ornstein-Uhlenbeck process

We now discuss the Ornstein-Uhlenbeck process for several variables which is described by the following drift and diffusion coefficients:

$$D_i(\{\mathbf{x}\}) = -\gamma_{ij}x_j \quad\text{and}\quad D_{ij}(\mathbf{x}) = D_{ij}, \tag{4.10}$$

where $\gamma_{ij}$ and $D_{ij}$ are constant matrices, and $D_{ij}$ is symmetric $D_{ij} = D_{ji}$. The corresponding Fokker-Planck equation for the transition probability is given by

$$\frac{\partial P\left(\{\mathbf{x}\},t|\{\mathbf{x}'\},t'\right)}{\partial t} = \left[\gamma_{ij}\frac{\partial}{\partial x_i}x_j + D_{ij}\frac{\partial^2}{\partial x_i x_j}\right]P\left(\{\mathbf{x}\},t|\{\mathbf{x}'\},t'\right). \tag{4.11}$$

Eq. (4.11) can be solved by the method of the Fourier transform. The pair of the Fourier transform is given by

$$P_k\left(\{\mathbf{k}\},t|\{\mathbf{x}'\},t'\right) = \int e^{-i\mathbf{k}\cdot\mathbf{x}}P\left(\{\mathbf{x}\},t|\{\mathbf{x}'\},t'\right)d^N x. \tag{4.12}$$

and

$$P\left(\{\mathbf{x}\},t|\{\mathbf{x}'\},t'\right) = \frac{1}{(2\pi)^N}\int e^{i\mathbf{k}\cdot\mathbf{x}}P_k\left(\{\mathbf{k}\},t|\{\mathbf{x}'\},t'\right)d^N k, \tag{4.13}$$

where $d^N x = dx_1 dx_2 ... dx_N$. Applying the Fourier transform to Eq. (4.11) we arrive at

$$\frac{\partial P_k\left(\{\mathbf{k}\}, t | \{\mathbf{x}'\}, t'\right)}{\partial t} = -\left[\gamma_{ij} k_i \frac{\partial}{\partial k_j} + D_{ij} k_i k_j\right] P_k\left(\{\mathbf{k}\}, t | \{\mathbf{x}'\}, t'\right).$$

(4.14)

We now seek a solution of the type [1]

$$P_k\left(\{\mathbf{k}\}, t | \{\mathbf{x}'\}, t'\right) = e^{-ik_n M_n(t-t') - \frac{k_n k_l \sigma_{nl}(t-t')}{2}},$$

(4.15)

with $\sigma_{ij}(t) = \sigma_{ji}(t)$. Substituting Eq. (4.15) into Eq. (4.14) we obtain

$$ik_n \frac{dM_n(t)}{dt} + \frac{k_n k_l}{2} \frac{d\sigma_{nl}(t)}{dt}$$

$$+ \gamma_{ij} k_i \left[i\delta_{nj} M_n(t) + \frac{\delta_{nj}}{2} k_l \sigma_{nl}(t) + \frac{\delta_{lj}}{2} k_n \sigma_{nl}(t)\right] - D_{ij} k_i k_j = 0.$$

(4.16)

Eq. (4.16) requires that the coefficients of $k_i$ and $k_i k_j$ vanish, i.e,

$$\frac{dM_i(t)}{dt} + \gamma_{ij} M_j(t) = 0$$

(4.17)

and

$$\frac{d\sigma_{ij}(t)}{dt} + \gamma_{il} \sigma_{lj}(t) + \gamma_{jl} \sigma_{li}(t) - 2D_{ij} = 0.$$

(4.18)

Note that in Eq. (4.18) the symmetry of the matrix $\gamma \sigma$ has been taken.

We now use the initial sharp condition

$$P\left(\{\mathbf{x}\}, t' | \{\mathbf{x}'\}, t'\right) = \delta\left(\{\mathbf{x}\} - \{\mathbf{x}'\}\right)$$

(4.19)

to determine the initial constants. In Fourier space we have

$$P_k\left(\{\mathbf{k}\}, t' | \{\mathbf{x}'\}, t'\right) = e^{-ik_j x'_j}.$$

(4.20)

Comparing Eq. (4.20) with Eq. (4.15) we obtain

$$M_n(t') = x'_n \quad \text{and} \quad \sigma_{nl}(0) = 0.$$

(4.21)

In matrix notation Eq. (4.17) is given by

$$\frac{dM(t)}{dt} + \gamma M(t) = 0.$$

(4.22)

Eq. (4.22) has the following formal solution:

$$M(t) = e^{-\gamma t} M(t'),$$

(4.23)

where

$$e^{-\gamma t} = I - \gamma t + \frac{1}{2} \gamma^2 t^2 + ...$$

(4.24)

and $I$ is the identity matrix. In terms of the matrix elements Eq. (4.23) is given by

$$M_i(t) = \left[e^{-\gamma t}\right]_{ij} x'_j. \tag{4.25}$$

We now consider a complete biorthogonal set for the matrix $\gamma$ [1] described by

$$\gamma_{ij} u_{j,\alpha} = \lambda_\alpha u_{i,\alpha} \tag{4.26}$$

and

$$v_{i,\alpha}\gamma_{ij} = \lambda_\alpha v_{j,\alpha}, \tag{4.27}$$

with the following orthonormality and completeness relation:

$$\sum_\alpha v_{i,\alpha}u_{j,\alpha} = \delta_{ij} \quad \text{and} \quad u_{i,\alpha}v_{i,\beta} = \delta_{\alpha\beta}. \tag{4.28}$$

It should be noted that the summation convention is only valid for the Latin indices. From Eq. (4.26) we have

$$\sum_\alpha \gamma_{ij}u_{j,\alpha}v_{k,\alpha} = \sum_\alpha \lambda_\alpha u_{i,\alpha}v_{k,\alpha}. \tag{4.29}$$

By using Eq. (4.28) we can reduce Eq. (4.29) to

$$\gamma_{ik} = \sum_\alpha \lambda_\alpha u_{i,\alpha}v_{k,\alpha}. \tag{4.30}$$

Now, using Eq. (4.30) we can write $\left[e^{-\gamma t}\right]_{ij}$ in terms of the biorthogonal set as follows.

$$\left[e^{-\gamma t}\right]_{ij} = \delta_{ij} - \gamma_{ij} + \frac{1}{2}\gamma_{ik}\gamma_{kj}t^2 + \dots. \tag{4.31}$$

Note that the product $\gamma^2$ can be written as follows:

$$\gamma_{ik}\gamma_{kj} = \sum_{\alpha,\beta} \lambda_\alpha u_{i,\alpha}v_{k,\alpha}\lambda_\beta u_{k,\beta}v_{j,\beta}; \tag{4.32}$$

using Eq. (4.28) we obtain

$$\gamma_{ik}\gamma_{kj} = \sum_{\alpha,\beta} \lambda_\alpha\lambda_\beta\delta_{\alpha\beta}u_{i,\alpha}v_{j,\beta} = \sum_\alpha \lambda_\alpha^2 u_{i,\alpha}v_{j,\alpha}. \tag{4.33}$$

For higher order, $\left[\gamma^n\right]_{ij}$ yields

$$\left[\gamma^n\right]_{ij} = \sum_\alpha \lambda_\alpha^n u_{i,\alpha}v_{j,\alpha}; \tag{4.34}$$

thus we have

$$\left[e^{-\gamma t}\right]_{ij} =$$

$$\sum_\alpha \left[ 1 - \lambda_\alpha t + \frac{1}{2}\lambda_\alpha^2 t^2 + ... \right] u_{i,\alpha} v_{j,\alpha} = \sum_\alpha e^{-\lambda_\alpha t} u_{i,\alpha} v_{j,\alpha}. \tag{4.35}$$

On the other hand, a solution of Eq. (4.18) can be obtained from the Langevin equation [1] and $\sigma_{ij}(t)$ corresponds to the covariance given by

$$\sigma_{ij}(t) = \int_0^t \left[ e^{-\gamma\tau} \right]_{ik} \left[ e^{-\gamma\tau} \right]_{jl} 2D_{kl} d\tau. \tag{4.36}$$

Substituting Eq. (4.35) into Eq. (4.36) we have

$$\sigma_{ij}(t) = 2\sum_{\alpha,\beta} \int_0^t e^{-(\lambda_\alpha - \lambda_\beta)\tau} u_{i,\alpha} v_{k,\alpha} u_{j,\beta} v_{l,\beta} D_{kl} d\tau. \tag{4.37}$$

Using the orthonormality and completeness relation (4.28) we obtain

$$\sigma_{ij}(t) = 2\sum_{\alpha,\beta} \frac{1 - e^{-(\lambda_\alpha + \lambda_\beta)t}}{\lambda_\alpha + \lambda_\beta} D_{\alpha,\beta} u_{i,\alpha} u_{j,\beta}, \tag{4.38}$$

where

$$D_{\alpha,\beta} = v_{k,\alpha} D_{kl} v_{l,\beta}. \tag{4.39}$$

The inverse Fourier transform of Eq. (4.15) is calculated as follows. Substituting Eq. (4.15) into Eq. (4.13) we have

$$P(\{\mathbf{x}\}, t | \{\mathbf{x}'\}, t') = \frac{1}{(2\pi)^N} \int e^{i(x_n - M_n)k_n - \frac{\sigma_{nl}k_n k_l}{2}} d^N k. \tag{4.40}$$

Note that the exponential of Eq. (4.40) can be written as

$$e^{i(x_n - M_n) - \frac{k_n k_l \sigma_{nl}}{2}} = e^{-\frac{q_n q_n}{2} - \frac{1}{2}\left(\sigma^{-1}\right)_{nl}(M_n - x_n)(M_l - x_l)} \tag{4.41}$$

with

$$q_n = \left(\sigma^{\frac{1}{2}}\right)_{nl} k_l + i\left(\sigma^{-\frac{1}{2}}\right)_{nl}(M_l - x_l). \tag{4.42}$$

The matrix $\sigma^{\frac{1}{2}}$ is the square root matrix of $\sigma$; it means that the matrix product $\sigma^{\frac{1}{2}}\sigma^{\frac{1}{2}}$ is equal to $\sigma$. Whereas $\sigma^{-1}$ is the inverse of $\sigma$, and $\sigma^{-\frac{1}{2}}$ is the inverse of $\sigma^{\frac{1}{2}}$. Substituting the variables $q_n$ into Eq. (4.40) and using Eq. (4.41) we arrive at

$$P(\{\mathbf{x}\}, t | \{\mathbf{x}'\}, t') = \frac{e^{-\frac{1}{2}\left(\sigma^{-1}\right)_{nl}(M_n - x_n)(M_l - x_l)}}{(2\pi)^N} \int e^{-\frac{q_n q_n}{2}} |J| d^N q, \tag{4.43}$$

where $J$ is the Jacobian of transformation [5] given by

$$J = \frac{\partial(k_1, k_2, ..., k_n)}{\partial(q_1, q_2, ..., q_n)} = \frac{1}{\frac{\partial(q_1, q_2, ..., q_n)}{\partial(k_1, k_2, ..., k_n)}}. \tag{4.44}$$

Using the relations

$$\frac{\partial q_i}{\partial k_j} = \left(\sigma^{\frac{1}{2}}\right)_{il} \delta_{lj} = \left(\sigma^{\frac{1}{2}}\right)_{ij} \tag{4.45}$$

and

$$\text{Det}\,(AB) = \text{Det}(A)\text{Det}(B) \tag{4.46}$$

we obtain from Eq. (4.44) the following result:

$$J = \frac{1}{\text{Det}\left(\sigma^{\frac{1}{2}}\right)_{ij}} = \frac{1}{\left(\text{Det}\,(\sigma)_{ij}\right)^{\frac{1}{2}}}. \tag{4.47}$$

Substituting Eq. (4.47) into (4.43) yields

$$P\left(\{\mathbf{x}\}, t | \{\mathbf{x}'\}, t'\right) = \frac{e^{-\frac{1}{2}\left(\sigma^{-1}\right)_{nl}(M_n - x_n)(M_l - x_l)}}{(2\pi)^N \left(\text{Det}\,(\sigma)_{ij}\right)^{\frac{1}{2}}} \int e^{-\frac{q_n q_n}{2}} d^N q$$

$$= \frac{e^{-\frac{1}{2}\left(\sigma^{-1}\right)_{nl}(M_n - x_n)(M_l - x_l)}}{(2\pi)^N \left(\text{Det}\,(\sigma)_{ij}\right)^{\frac{1}{2}}} \left(\int_{-\infty}^{\infty} e^{-\frac{q^2}{2}} dq\right)^N$$

$$= \frac{e^{-\frac{1}{2}\left(\sigma^{-1}(t-t')\right)_{nl}\left(M_n(t-t') - x_n\right)\left(M_l(t-t') - x_l\right)}}{(2\pi)^{\frac{N}{2}} \left(\text{Det}\,(\sigma\,(t - t'))_{ij}\right)^{\frac{1}{2}}}. \tag{4.48}$$

Eq. (4.48) is the transition probability for the Ornstein-Uhlenbeck process with several variables.

Note that a stationary solution for the transition probability (4.48) can be obtained if all real parts of the eigenvalues of matrix $\gamma_{ij}$ are positive; in this case for large $(t - t')$ the matrix elements $M_i$ tend to zeros and the transition probability (4.48) reduces to

$$\rho_{st}\left(\{\mathbf{x}\}\right) = \frac{e^{-\frac{1}{2}\left(\sigma^{-1}(\infty)\right)_{nl} x_n x_l}}{(2\pi)^{\frac{N}{2}} \left(\text{Det}\,(\sigma\,(\infty))_{ij}\right)^{\frac{1}{2}}}. \tag{4.49}$$

The matrix elements $\sigma_{ij}\,(\infty)$ can be obtained from Eq. (4.38), and they are given by

$$\sigma_{ij}\,(\infty) = 2 \sum_{\alpha,\beta} \frac{D_{\alpha,\beta} u_{i,\alpha} u_{j,\beta}}{\lambda_\alpha + \lambda_\beta}. \tag{4.50}$$

From the the transition probability (4.48) and the stationary distribution (4.49) we may form the joint probability density $P\left(\{\mathbf{x}\}, t | \{\mathbf{x}'\}, t'\right) \rho_{st}\left(\{\mathbf{x}\}\right)$ which can be used to calculate any two-time expectation value.

## 4.3 The Klein-Kramers equation for a linear force in a one-dimensional space

The Klein-Kramers equation, for a one-dimensional space, describing the Brownian motion of particles with mass $m$ subjected to an external force is described by the following Langevin equation:

$$\frac{d^2x}{dt^2} + \gamma\frac{dx}{dt} - \frac{F(x)}{m} = L(t) \tag{4.51}$$

with $L(t)$ Gaussian distributed given by

$$\langle L(t)\rangle = 0 \quad \text{and} \quad \langle L(t)L(t')\rangle = 2\gamma v_{th}^2\delta(t - t'), \tag{4.52}$$

where $v_{th}^2 = k_BT/m$; the corresponding Fokker-Planck equation is given by

$$\frac{\partial\rho}{\partial t} = \left[-\frac{\partial}{\partial x}v + \frac{\partial}{\partial v}\left(\gamma v - \frac{F(x)}{m}\right) + \gamma v_{th}^2\frac{\partial^2}{\partial v^2}\right]\rho. \tag{4.53}$$

Note that the difference between the Smoluchowski and Klein-Kramers equations is the inertial term $d^2x/dt^2$ which appears in the Langevin equation (4.51). For large friction limit the inertial term can be omitted and the Klein-Kramers equation reduces to the Smoluchowski equation.

In the case of linear force we have

$$\frac{\partial\rho}{\partial t} = \left[-\frac{\partial}{\partial x}v + \frac{\partial}{\partial v}\left(\gamma v - \omega^2 x\right) + \gamma v_{th}^2\frac{\partial^2}{\partial v^2}\right]\rho, \tag{4.54}$$

Now, a solution for Eq. (4.51), in the case of linear force, can be obtained from the general solution of the Ornstein-Uhlenbeck process described in the previous section. Rewrite Eq. (4.51), with $F(x) = -m\omega^2 x$, as follows:

$$\frac{d}{dt}\begin{pmatrix} x \\ v \end{pmatrix} = -\begin{pmatrix} 0 & -1 \\ \omega^2 & \gamma \end{pmatrix}\begin{pmatrix} x \\ v \end{pmatrix} + \begin{pmatrix} 0 \\ L(t) \end{pmatrix} \tag{4.55}$$

or

$$\frac{dx_i}{dt} = -\bar{\gamma}_{ij}x_j + L_i(t), \tag{4.56}$$

where the matrices $\bar{\gamma}$ and $D$ are given by

$$\bar{\gamma} = \begin{pmatrix} 0 & -1 \\ \omega^2 & \gamma \end{pmatrix} \tag{4.57}$$

and

$$D = \begin{pmatrix} 0 & 0 \\ 0 & \gamma v_{th}^2 \end{pmatrix}. \tag{4.58}$$

Using the relations (4.26)-(4.30) we obtain

$$u_{1,1}v_{1,1} = \frac{\lambda_2}{\lambda_2 - \lambda_1}, \quad u_{1,2}v_{1,2} = -\frac{\lambda_1}{\lambda_2 - \lambda_1}, \tag{4.59}$$

$$u_{2,\alpha} = -\lambda_\alpha u_{1,\alpha}, \quad v_{2,\alpha} = \frac{\lambda_\alpha}{\omega^2}v_{1,\alpha}, \tag{4.60}$$

$$u_{2,2}v_{2,2} = \frac{\lambda_2}{\lambda_2 - \lambda_1}, \quad u_{2,1}v_{2,1} = \frac{1}{\lambda_1}\left[\gamma - \frac{\lambda_2^2}{\lambda_2 - \lambda_1}\right], \tag{4.61}$$

$$\lambda_\alpha = \frac{\gamma \pm \sqrt{\gamma^2 - 4\omega^2}}{2} \quad \text{and} \quad \omega^2 = \lambda_1\lambda_2. \tag{4.62}$$

Note that the relations (4.62) are satisfied only if the signs of $\lambda_1$ and $\lambda_2$ are opposed, which imply that

$$\lambda_1 + \lambda_2 = \gamma. \tag{4.63}$$

The eigenvectors $u_{i,\alpha}$ and $v_{i,\alpha}$ are given by

$$\mathbf{u}_{\alpha=1} = u_{1,1}\begin{pmatrix} 1 \\ -\lambda_1 \end{pmatrix}, \quad \mathbf{u}_{\alpha=2} = u_{1,2}\begin{pmatrix} 1 \\ -\lambda_2, \end{pmatrix} \tag{4.64}$$

$$\mathbf{v}_{\alpha=1} = \frac{(\lambda_2, 1)}{u_{1,1}(\lambda_2 - \lambda_1)} \quad \text{and} \quad \mathbf{v}_{\alpha=2} = -\frac{(\lambda_1, 1)}{u_{1,2}(\lambda_2 - \lambda_1)}. \tag{4.65}$$

The expression (4.35) can be determined by using the eigenvectors (4.64) and (4.65), with $u_{1,1} = -1$ and $u_{1,2} = 1$, and it yields

$$\left[e^{-\bar{\gamma}t}\right]_{11} = \frac{\lambda_1 e^{-\lambda_2 t} - \lambda_2 e^{-\lambda_1 t}}{\lambda_1 - \lambda_2}, \quad \left[e^{-\bar{\gamma}t}\right]_{12} = \frac{e^{-\lambda_2 t} - e^{-\lambda_1 t}}{\lambda_1 - \lambda_2}, \tag{4.66}$$

$$\left[e^{-\bar{\gamma}t}\right]_{21} = \omega^2 \frac{e^{-\lambda_1 t} - e^{-\lambda_2 t}}{\lambda_1 - \lambda_2}, \quad \text{and} \quad \left[e^{-\bar{\gamma}t}\right]_{22} = \frac{\lambda_1 e^{-\lambda_1 t} - \lambda_2 e^{-\lambda_2 t}}{\lambda_1 - \lambda_2}, \tag{4.67}$$

From the symmetric matrix $\sigma$ and the relation $\sigma^{-1}\sigma = I$ we obtain

$$\sigma_{11}^{-1}\sigma_{11} + \sigma_{12}^{-1}\sigma_{21} = 1, \tag{4.68}$$

$$\sigma_{11}^{-1}\sigma_{12} + \sigma_{12}^{-1}\sigma_{22} = 0 \rightarrow \sigma_{12}^{-1} = -\frac{\sigma_{11}^{-1}\sigma_{12}}{\sigma_{22}}, \tag{4.69}$$

$$\sigma_{21}^{-1}\sigma_{11} + \sigma_{22}^{-1}\sigma_{12} = 0 \tag{4.70}$$

and

$$\sigma_{21}^{-1}\sigma_{12} + \sigma_{22}^{-1}\sigma_{22} = 1. \tag{4.71}$$

Substituting Eq. (4.69) into Eq. (4.68) yields

$$\sigma_{11}^{-1} = \frac{\sigma_{22}}{\text{Det}(\sigma)}. \tag{4.72}$$

The other components of the matrix $\sigma^{-1}$ are given by

$$\sigma_{12}^{-1} = \sigma_{21}^{-1} = -\frac{\sigma_{12}}{\text{Det}(\sigma)}, \tag{4.73}$$

$$\sigma_{22}^{-1} = \frac{\sigma_{11}}{\text{Det}(\sigma)} \tag{4.74}$$

and

$$\text{Det}(\sigma) = \sigma_{11}\sigma_{22} - (\sigma_{12})^2. \tag{4.75}$$

Using the components of the eigenvectors (4.64) and (4.65) and Eq. (4.38) we obtain the following explicit expressions for $\sigma_{ij}$:

$$\sigma_{11} = \frac{\gamma v_{th}^2}{(\lambda_1 - \lambda_2)^2} \times$$

$$\left[ \frac{\lambda_1 + \lambda_2}{\lambda_1 \lambda_2} - 4 \frac{1 - e^{-(\lambda_1 + \lambda_2)t}}{\lambda_1 + \lambda_2} - \frac{e^{-2\lambda_1 t}}{\lambda_1} - \frac{e^{-2\lambda_2 t}}{\lambda_2} \right], \tag{4.76}$$

$$\sigma_{12} = \sigma_{21} = \frac{\gamma v_{th}^2}{(\lambda_1 - \lambda_2)^2} \left( e^{-\lambda_1 t} - e^{-\lambda_2 t} \right)^2 \tag{4.77}$$

and

$$\sigma_{22} = \frac{\gamma v_{th}^2}{(\lambda_1 - \lambda_2)^2}$$

$$\times \left[ \lambda_1 + \lambda_2 + \frac{4\lambda_1 \lambda_2}{\lambda_1 + \lambda_2} \left( e^{-(\lambda_1 + \lambda_2)t} - 1 \right) - \lambda_1 e^{-2\lambda_1 t} - \lambda_2 e^{-2\lambda_2 t} \right]. \tag{4.78}$$

We can see that, from Eq. (4.62), the real parts of the eigenvalues $\lambda_\alpha$ are positive for $\gamma > 0$. Therefore the stationary solution exists. For $t \to \infty$, we obtain from Eqs. (4.66)-(4.78) the following expressions:

$$\left[ e^{-\bar{\gamma} t} \right]_{ij} = 0, \tag{4.79}$$

$$\sigma_{11} = \frac{v_{th}^2}{\lambda_1 \lambda_2} = \frac{v_{th}^2}{\omega^2}, \quad \sigma_{12} = \sigma_{21}(t \to \infty) = 0, \quad \sigma_{22} = v_{th}^2, \tag{4.80}$$

$$\text{Det}(\sigma) = \frac{v_{th}^4}{\omega^2}, \quad \sigma_{11}^{-1} = \frac{\omega^2}{v_{th}^2}, \quad \sigma_{12}^{-1} = \sigma_{21}^{-1} = 0 \text{ and } \sigma_{22}^{-1} = \frac{1}{v_{th}^2}. \tag{4.81}$$

Substituting the results (4.80)-(4.81) into (4.49) yields

$$\rho_{st}(x,v) = \frac{\omega}{2\pi v_{th}^2} e^{-\frac{1}{K_B T}\left(\frac{m\omega^2 x^2 + mv^2}{2}\right)} = \frac{\omega}{2\pi v_{th}^2} e^{-\frac{E}{K_B T}}, \qquad (4.82)$$

where $E$ is the mechanical energy given by

$$E = \frac{m\omega^2 x^2}{2} + \frac{mv^2}{2}. \qquad (4.83)$$

Eq. (4.82) shows the Boltzmann distribution.

For the force-free case $\omega \to 0$, the eigenvalues $\lambda_\alpha$ are given by

$$\lambda_\alpha = \frac{\gamma \pm \gamma}{2} \quad \text{and} \quad \lambda_1 + \lambda_2 = \gamma. \qquad (4.84)$$

We may choose $\lambda_1 = 0$ and $\lambda_2 = \gamma$ or $\lambda_1 = \gamma$ and $\lambda_2 = 0$. The two sets of eigenvalues give the same result for the transition probability. The expressions (4.66)-(4.67) and (4.76)-(4.78) reduce to

$$\left[e^{-\bar{\gamma}t}\right]_{11} = 1, \quad \left[e^{-\bar{\gamma}t}\right]_{12} = \frac{1 - e^{-\gamma t}}{\gamma}, \qquad (4.85)$$

$$\left[e^{-\bar{\gamma}t}\right]_{21} = 0, \quad \left[e^{-\bar{\gamma}t}\right]_{22} = e^{-\gamma t}, \qquad (4.86)$$

$$\sigma_{11} = \frac{v_{th}^2}{\gamma}\left(2t - \frac{3 - 4e^{-\gamma t} + e^{-2\gamma t}}{\gamma}\right), \qquad (4.87)$$

$$\sigma_{12} = \sigma_{21} = \frac{v_{th}^2}{\gamma}\left(1 - e^{-\gamma t}\right)^2, \qquad (4.88)$$

and

$$\sigma_{22} = v_{th}^2\left(1 - e^{-2\gamma t}\right). \qquad (4.89)$$

Therefore, the result for the transition probability is given by

$$P(x,v,t|x',v',0)$$

$$= \frac{e^{-\frac{1}{2\mathrm{Det}(\sigma(t))}\left[-\sigma_{22}(x-x(t))^2 + \sigma_{12}(x-x(t))(v-v(t)) - \sigma_{11}(v-v(t))^2\right]}}{(2\pi)(\mathrm{Det}(\sigma(t)))^{\frac{1}{2}}}, \qquad (4.90)$$

where $x(t)$ and $v(t)$ are given by

$$x(t) = x' + v'\frac{1 - e^{-\gamma t}}{\gamma}, \qquad (4.91)$$

and

$$v(t) = v'e^{-\gamma t}. \qquad (4.92)$$

Fig. 4.1 shows the transition probability (4.90) for $v_{th}^2 = 1$ and $t = 0.9$.

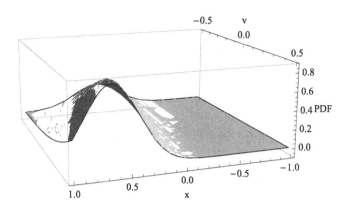

Fig. 4.1 Plots of the transition probability, Eq. (4.90), for $v_{th}^2 = 1$ and $t = 0.9$. The initial values are: $x' = 0$ and $v' = 1$.

## 4.4 Harmonic oscillator driven by colored noise in a one-dimensional space

The colored noise has been applied to nonlinear dynamics [6–9]. As an application we consider the underdamped harmonic oscillator driven by the harmonic noise, for a one-dimensional space. The equation is described by

$$\frac{d^2x}{dt^2} + \gamma\frac{dx}{dt} + \omega_0^2 x = y(t),\tag{4.93}$$

where $y(t)$ is the harmonic noise (see Appendix 4.6.1) which obeys the following Langevin equation:

$$\frac{d^2y}{dt^2} + \Gamma\frac{dy}{dt} + \Omega_0^2 y = L(t),\tag{4.94}$$

$$\langle L(t)\rangle = 0 \quad \text{and} \quad \langle L(t)L(t')\rangle = 2\epsilon\Omega_0^4\delta(t - t'),\tag{4.95}$$

where $\gamma$, $\Gamma$ and $\epsilon$ are positive constants. Note that the harmonic noise may be associated with an RLC circuit (combination of resistance R, inductance L and capacitance C) driven by the white noise.

Eqs. (4.93) and (4.94) can be written in terms of the first derivatives as follows:

$$\frac{dx}{dt} = v,\tag{4.96}$$

$$\frac{dv}{dt} + \gamma v + \omega_0^2 x = y(t),\tag{4.97}$$

$$\frac{dy}{dt} = s, \tag{4.98}$$

$$\frac{ds}{dt} + \Gamma s + \Omega_0^2 y = L(t). \tag{4.99}$$

The corresponding Fokker-Planck equation is obtained as follows. First we identify $\xi_1 = x$, $\xi_2 = v$, $\xi_3 = y$ and $\xi_4 = s$. The coefficients $D_i$ and $D_{ij}$ are obtained from Eqs. (2.52)-(2.53). The coefficients $D_i$ are given by

$$D_1 = v, D_2 = -\gamma v - \omega_0^2 x + y, D_3 = s, D_4 = -\Gamma s - \Omega_0^2 y. \tag{4.100}$$

Whereas, the $D_{ij}$ are zeros except $D_{44} = \epsilon \Omega_0^4$. Substituting the coefficients $D_i$ and $D_{ij}$ into Eq. (2.55) we obtain the following Fokker-Planck equation:

$$\frac{\partial \rho}{\partial t} = \left[ -v \frac{\partial}{\partial x} \right.$$

$$\left. + \frac{\partial}{\partial v} \left( \gamma v + \omega_0^2 x - y \right) - s \frac{\partial}{\partial y} + \frac{\partial}{\partial s} \left( \Gamma s + \Omega_0^2 y \right) + \epsilon \Omega_0^4 \frac{\partial^2}{\partial s^2} \right] \rho. \tag{4.101}$$

Eq. (4.101) contains five variables $(t, x, v, y, s)$ and four-dimensional state space.

### 4.4.1    *Stationary solution*

Fortunately, the stationary solution of Eq. (4.101) can be determined. The stationary state is obtained by setting $\partial \rho / \partial t = 0$, and the Fourier transform of Eq. (4.101) is given by

$$\left[ (k_x - \gamma k_v) \frac{\partial}{\partial k_v} - \omega_0^2 k_v \frac{\partial}{\partial k_x} \right.$$

$$\left. + \left( k_v - \Omega_0^2 k_s \right) \frac{\partial}{\partial k_y} + (k_y - \Gamma k_s) \frac{\partial}{\partial k_s} \right] \rho_k = \epsilon \Omega_0^4 k_s^2 \rho_k. \tag{4.102}$$

Now, we seek a solution of the type [1]

$$\rho_k \left( \{\mathbf{k}\} \mid \{\mathbf{x}'\} \right) = e^{-ik_n M_n - \frac{k_n k_l \sigma_{nl}}{2}}, \tag{4.103}$$

with $\sigma_{ij} = \sigma_{ji}$. Setting $k_1 = k_x$, $k_2 = k_v$, $k_3 = k_y$ and $k_4 = k_s$ we obtain the following solutions for the coefficients $M_n$ and $\sigma_{ij}$:

$$M_1 = M_2 = M_3 = M_4 = 0, \tag{4.104}$$

$$\sigma_{12} = \sigma_{34} = 0, \tag{4.105}$$

$$\sigma_{11} = \frac{\epsilon \left( \Gamma + \frac{\Omega_0^2}{\gamma} + \frac{\omega_0^2 - \Omega_0^2}{\gamma + \Gamma} \right)}{\omega_0^2 \Gamma \left[ \gamma + \frac{\Gamma \omega_0^2}{\Omega_0^2} + \frac{\left( \omega_0^2 - \Omega_0^2 \right)^2}{\Omega_0^2 (\gamma + \Gamma)} \right]}, \tag{4.106}$$

$$\sigma_{13} = \frac{\epsilon \left[ \Gamma (\gamma + \Gamma) + \omega_0^2 - \Omega_0^2 \right]}{\Gamma (\gamma + \Gamma) \left[ \gamma + \frac{\Gamma \omega_0^2}{\Omega_0^2} + \frac{\left( \omega_0^2 - \Omega_0^2 \right)^2}{\Omega_0^2 (\gamma + \Gamma)} \right]}, \tag{4.107}$$

$$\sigma_{23} = \gamma \sigma_{22} = -\sigma_{14} = \frac{\epsilon \Omega_0^2}{\Gamma \left[ \gamma + \frac{\Gamma \omega_0^2}{\Omega_0^2} + \frac{\left( \omega_0^2 - \Omega_0^2 \right)^2}{\Omega_0^2 (\gamma + \Gamma)} \right]}, \tag{4.108}$$

$$\sigma_{24} = \frac{\omega_0^2}{\gamma + \Gamma} \frac{\Omega_0^2}{} \sigma_{23}, \quad \sigma_{33} = \frac{\epsilon \Omega_0^2}{\Gamma}, \quad \sigma_{44} = \frac{\epsilon \Omega_0^4}{\Gamma}, \tag{4.109}$$

and

$$\rho_k (k_x, k_v, k_y, k_s) = \exp \left[ -\frac{1}{2} \left( \sigma_{11} k_x^2 + \sigma_{22} k_v^2 + \sigma_{33} k_y^2 + \sigma_{44} k_s^2 \right) \right.$$

$$\left. - \left( \sigma_{13} k_x k_y + \sigma_{14} k_x k_s + \sigma_{23} k_v k_y + \sigma_{24} k_v k_s \right) \right]. \tag{4.110}$$

The inverse Fourier transform of Eq. (4.110) is given by

$$\rho_{st} = \frac{1}{(2\pi)^4} \int \exp \left[ i \left( k_x x + k_v v + k_y y + k_s s \right) \right.$$

$$- \frac{1}{2} \left( \sigma_{11} k_x^2 + \sigma_{22} k_v^2 + \sigma_{33} k_y^2 + \sigma_{44} k_s^2 \right)$$

$$\left. - \left( \sigma_{13} k_x k_y + \sigma_{14} k_x k_s + \sigma_{23} k_v k_y + \sigma_{24} k_v k_s \right) \right] dk_x dk_v dk_y dk_s. \tag{4.111}$$

Integration of Eq. (4.111) yields

$$\rho_{st} (x, v, y, s) = \frac{e^{-\frac{1}{2} \left( \frac{x^2}{\langle x^2 \rangle} + \frac{v^2}{\langle v^2 \rangle} + \frac{w^2}{A} + \frac{q^2}{B} \right)}}{(2\pi)^2 \sqrt{\langle x^2 \rangle \langle v^2 \rangle AB}}, \tag{4.112}$$

where

$$q = \gamma v + \omega_0^2 \left[ 1 - \frac{1}{1 + \frac{\gamma \Gamma}{\Omega_0^2} + \frac{\gamma \left( \omega_0^2 - \Omega_0^2 \right)}{\Omega_0^2 (\gamma + \Gamma)}} \right] x - y, \tag{4.113}$$

$$w = s - \gamma y + \frac{\gamma \left[ \gamma (\gamma + \Gamma) - \omega_0^2 + \Omega_0^2 \right]}{\gamma + \Gamma} v + \gamma \omega_0^2 x, \tag{4.114}$$

$$\langle x^2 \rangle = \sigma_{11}, \tag{4.115}$$

$$\langle v^2 \rangle = \sigma_{22}. \tag{4.116}$$

$$A = \frac{\epsilon \Omega_0^4}{(\gamma + \Gamma)}. \tag{4.117}$$

and

$$B = \frac{\epsilon \Omega_0^4}{\gamma \Gamma (\gamma + \Gamma) + \gamma \omega_0^2 + \Gamma \Omega_0^2}. \tag{4.118}$$

Eq. (4.112) is the stationary solution and it has a Gaussian form; besides, it is independent of the initial distribution. Note that $\sigma_{11}$, $\sigma_{22}$, $A$ and $B$ have positive values which guarantee that the PDF decays to zero for large values of $x$, $v$, $y$ and $s$.

From the solution (4.112) we can calculate $\langle y^2 \rangle$, $\langle s^2 \rangle$ and the mean energies [7,10] of the driving harmonic noise and linear oscillator which are given by

$$\langle y^2 \rangle = \sigma_{33} = \frac{\epsilon \Omega_0^2}{\Gamma}, \tag{4.119}$$

$$\langle s^2 \rangle = \sigma_{44} = \frac{\epsilon \Omega_0^4}{\Gamma}, \tag{4.120}$$

$$\langle E_{HN} \rangle = \frac{\langle s^2 \rangle}{2} + \Omega_0^2 \frac{\langle y^2 \rangle}{2} = \frac{\epsilon \Omega_0^4}{\Gamma} \tag{4.121}$$

and

$$\langle E_{HO} \rangle = \frac{\langle v^2 \rangle}{2} + \omega_0^2 \frac{\langle x^2 \rangle}{2} = \frac{\sigma_{22} + \omega_0^2 \sigma_{11}}{2} =$$

$$= \frac{\epsilon \left[ 2\frac{\Omega_0^2}{\gamma} + \Gamma + \frac{\omega_0^2 - \Omega_0^2}{\gamma + \Gamma} \right]}{2\Gamma \left[ \gamma + \frac{\Gamma \omega_0^2}{\Omega_0^2} + \frac{(\omega_0^2 - \Omega_0^2)^2}{\Omega_0^2 (\gamma + \Gamma)} \right]}. \tag{4.122}$$

Fig. 4.2 shows the mean energy $\langle E_{HO} \rangle$, Eq. (4.122), versus the frequency of the driving oscillator $\omega_0$. One can see the mean energy shows a peak at $\omega_0 = 1.396$ which is calculated by

$$\omega_0 = \Omega_0 \sqrt{1 - (\gamma + \Gamma) \left( \frac{2}{\gamma} + \frac{\Gamma}{\Omega_0^2} \right) \left[ 1 - \sqrt{1 + \frac{1 - \Gamma \left( \frac{2}{\gamma} + \frac{\Gamma}{\Omega_0^2} \right)}{\Omega_0^2 \left( \frac{2}{\gamma} + \frac{\Gamma}{\Omega_0^2} \right)^2}} \right]} \tag{4.123}$$

The peak is due to the resonance.

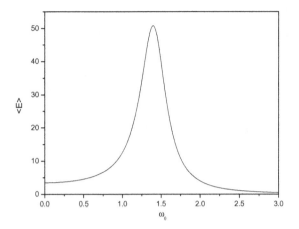

Fig. 4.2    Plot of the mean energy $\langle E_{HO} \rangle$ versus the frequency of the driving oscillator $\omega_0$, Eq. (4.122), for $\epsilon = 0.3$, $\gamma = 0.1$, $\Gamma = 0.3$ and $\Omega_0^2 = 2.0$.

### 4.4.2    *Time-dependent solution*

The time-dependent solution of Eq. (4.101) can also be obtained by using the method of the Fourier transform. In this case the initial condition should be given, and we assume a sharp condition described by

$$\rho\left(\{\mathbf{x}\}, t' \middle| \{\mathbf{x}'\}, t'\right) = \delta\left(\{\mathbf{x}\} - \{\mathbf{x}'\}\right); \tag{4.124}$$

in Fourier space it is given by

$$\rho_k\left(\{\mathbf{k}\}, t' \middle| \{\mathbf{x}'\}, t'\right) = e^{-ik_j x'_j}. \tag{4.125}$$

Applying the Fourier transform to Eq. (4.101) we arrive at

$$\frac{\partial \rho_k}{\partial t} = \left[ (k_x - \gamma k_v) \frac{\partial}{\partial k_v} - \omega_0^2 k_v \frac{\partial}{\partial k_x} \right.$$

$$\left. + \left(k_v - \Omega_0^2 k_s\right) \frac{\partial}{\partial k_y} + (k_y - \Gamma k_s) \frac{\partial}{\partial k_s} - \epsilon \Omega_0^4 k_s^2 \right] \rho_k. \tag{4.126}$$

Now, we seek a solution of the type [1]

$$\rho_k\left(\{\mathbf{k}\}, t \middle| \{\mathbf{x}'\}, t'\right) = e^{-ik_n M_n(t) - \frac{k_n k_l \sigma_{nl}(t)}{2}}, \tag{4.127}$$

with $\sigma_{ij}(t) = \sigma_{ji}(t)$. Comparing Eq. (4.125) with Eq. (4.127) yields

$$M_n(t') = x'_n \quad \text{and} \quad \sigma_{nl}(t') = 0. \tag{4.128}$$

Substituting Eq. (4.127) into Eq. (4.126) gives

$$i \left[ k_n \frac{dM_n(t)}{dt} - M_2(t)k_x + \left( \gamma M_2(t) + \omega_0^2 M_1(t) - M_3(t) \right) k_v - M_4(t)k_y \right.$$

$$\left. + \left( \Gamma M_4(t) + \Omega_0^2 M_3(t) \right) k_s \right] + \frac{d\sigma_{nl}(t)}{dt} \frac{k_n k_l}{2} - \sigma_{2l}(t)k_x k_l$$

$$+ \left( \gamma \sigma_{2l}(t) + \omega_0^2 \sigma_{1l}(t) - \sigma_{3l}(t) \right) k_v k_l - \sigma_{4l}(t)k_y k_l$$

$$+ \left( \Gamma \sigma_{4l}(t) + \Omega_0^2 \sigma_{3l}(t) \right) k_s k_l - \epsilon \Omega_0^4 k_s^2 = 0. \tag{4.129}$$

Eq. (4.129) requires that the coefficients of $k_i$ and $k_i k_j$ vanish, and the following differential equations are obtained:

$$\frac{dM_1}{dt} = M_2 \quad , \quad \frac{dM_2}{dt} + \gamma M_2 + \omega_0^2 M_1 - M_3 = 0, \tag{4.130}$$

$$\frac{dM_3}{dt} = M_4 \quad , \quad \frac{dM_4}{dt} + \Gamma M_4 + \Omega_0^2 M_3 = 0, \tag{4.131}$$

$$\frac{d\sigma_{11}}{dt} = 2\sigma_{12} \quad , \quad \frac{d\sigma_{12}}{dt} + \gamma \sigma_{12} - \sigma_{22} - \sigma_{13} + \omega_0^2 \sigma_{11} = 0, \tag{4.132}$$

$$\frac{d\sigma_{13}}{dt} = \sigma_{23} + \sigma_{14} \quad , \quad \frac{d\sigma_{14}}{dt} + \Gamma \sigma_{14} - \sigma_{24} + \Omega_0^2 \sigma_{13} = 0, \tag{4.133}$$

$$\frac{d\sigma_{22}}{dt} + 2 \left( \gamma \sigma_{22} - \sigma_{23} + \omega_0^2 \sigma_{12} \right) = 0 \quad , \tag{4.134}$$

$$\frac{d\sigma_{23}}{dt} + \gamma \sigma_{23} - \sigma_{24} - \sigma_{33} + \omega_0^2 \sigma_{13} = 0, \tag{4.135}$$

$$\frac{d\sigma_{24}}{dt} + (\gamma + \Gamma)\sigma_{24} - \sigma_{34} + \omega_0^2 \sigma_{14} + \Omega_0^2 \sigma_{23} = 0 \quad , \quad \frac{d\sigma_{33}}{dt} = 2\sigma_{34}, \tag{4.136}$$

$$\frac{d\sigma_{34}}{dt} + \Gamma \sigma_{34} - \sigma_{44} + \Omega_0^2 \sigma_{33} = 0 \tag{4.137}$$

and

$$\frac{d\sigma_{44}}{dt} + 2 \left( \Gamma \sigma_{44} + \Omega_0^2 \sigma_{34} - \epsilon \Omega_0^4 \right) = 0. \tag{4.138}$$

Now, we consider that the driving noise is zero-centered $\langle y(t) \rangle = 0$ which implies that

$$M_3(t) = M_4(t) = 0. \tag{4.139}$$

With the result (4.139) we can solve Eq. (4.130) easily, and the coefficients $M_1(t)$ and $M_2(t)$ have the following solutions:

$$M_1(t) = \begin{cases} \left[ x_0 \cosh(\bar{\omega}_1 t) + (v_0 + \gamma x_0) \frac{\sinh(\bar{\omega}_1 t)}{\bar{\omega}_1} \right] e^{-\frac{\gamma}{2}t} & \text{if } \omega_0^2 < \frac{\gamma^2}{4} \\ \left[ x_0 + \left( v_0 + \frac{\gamma}{2} x_0 \right) t \right] e^{-\frac{\gamma}{2}t} & \text{if } \omega_0^2 = \frac{\gamma^2}{4} \\ \left[ x_0 \cos(\bar{\omega}_2 t) + (v_0 + \gamma x_0) \frac{\sin(\bar{\omega}_2 t)}{\bar{\omega}_2} \right] e^{-\frac{\gamma}{2}t} & \text{if } \omega_0^2 > \frac{\gamma^2}{4} \end{cases} \qquad (4.140)$$

and

$$M_2(t) = \begin{cases} \left[ v_0 \cosh(\bar{\omega}_1 t) - \omega_0^2 x_0 \frac{\sinh(\bar{\omega}_1 t)}{\bar{\omega}_1} \right] e^{-\frac{\gamma}{2}t} & \text{if } \omega_0^2 < \frac{\gamma^2}{4} \\ \left[ v_0 - \left( \omega_0^2 x_0 + \frac{\gamma}{2} v_0 \right) t \right] e^{-\frac{\gamma}{2}t} & \text{if } \omega_0^2 = \frac{\gamma^2}{4} \\ \left[ v_0 \cos(\bar{\omega}_2 t) - \omega_0^2 x_0 \frac{\sin(\bar{\omega}_2 t)}{\bar{\omega}_2} \right] e^{-\frac{\gamma}{2}t} & \text{if } \omega_0^2 > \frac{\gamma^2}{4} \end{cases}, \qquad (4.141)$$

where

$$\bar{\omega}_1 = \sqrt{\frac{\gamma^2}{4} - \omega_0^2} \qquad (4.142)$$

and

$$\bar{\omega}_2 = \sqrt{\omega_0^2 - \frac{\gamma^2}{4}}. \qquad (4.143)$$

However, the coefficients $\sigma_{ij}$ have not simple solutions. In this case they are determined numerically; first, they are transformed to Laplace space, then the results are obtained by using a numerical algorithm for inversion of the Laplace transform [11] (see Appendix). In Laplace space, the coefficients $\sigma_{ij}$ are given by

$$\bar{\sigma}_{33}(u) = \frac{4\epsilon \Omega_0^4}{u \left[ 2\Omega_0^2 u + (2\Gamma + u)(u^2 + \Gamma u + 2\Omega_0^2) \right]}, \quad \bar{\sigma}_{34}(u) = \frac{u \bar{\sigma}_{33}(u)}{2}, \qquad (4.144)$$

$$\bar{\sigma}_{44}(u) = \frac{2}{(2\Gamma + u)} \left( \frac{\epsilon \Omega_0^4}{u} - \frac{\Omega_0^2 u}{2} \bar{\sigma}_{33}(u) \right), \qquad (4.145)$$

$$\bar{\sigma}_{14}(u) = \frac{\left[ u(u^2 + \gamma u + \omega_0^2 - \Omega_0^2) - 2\Omega_0^2(\gamma + \Gamma + 2u) \right] \bar{\sigma}_{34}}{g_1(u) + g_2(u)}, \qquad (4.146)$$

$$\bar{\sigma}_{23}(u) = \frac{(u^2 + \Gamma u + \Omega_0^2 - \omega_0^2)\bar{\sigma}_{14}(u) + u \bar{\sigma}_{33}(u)}{u^2 + \gamma u + \omega_0^2 - \Omega_0^2}, \qquad (4.147)$$

$$\bar{\sigma}_{13}(u) = \frac{\bar{\sigma}_{14}(u) + \bar{\sigma}_{23}(u)}{u}, \qquad (4.148)$$

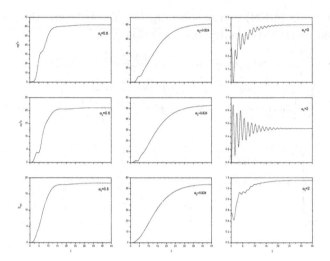

Fig. 4.3    Plots of the second moments for the position and velocity ($\langle x^2 \rangle$ and $\langle v^2 \rangle$), and the mean energy $\langle E_{HO} \rangle$ in function of time for different values of $\omega_0$. The initial values and the parameters are given by $x_0 = 0.5$, $v_0 = 0.2$, $\epsilon = 1.2$, $\gamma = 0.2$, $\Gamma = 0.2$ and $\Omega_0^2 = 0.7$.

$$\bar{\sigma}_{24}(u) = \frac{-\omega_0^2 \bar{\sigma}_{14}(u) - \Omega_0^2 \bar{\sigma}_{23} + \bar{\sigma}_{34}(u)}{\gamma + \Gamma + u}, \tag{4.149}$$

$$\bar{\sigma}_{22}(u) = \frac{2\left[(u^2 + \gamma u + 2\omega_0^2)\,\bar{\sigma}_{23}(u) - \omega_0^2 u \bar{\sigma}_{13}(u)\right]}{2\omega_0^2 u + (2\gamma + u)(u^2 + \gamma u + 2\omega_0^2)}, \tag{4.150}$$

$$\bar{\sigma}_{11}(u) = \frac{2\,(\bar{\sigma}_{13}(u) + \bar{\sigma}_{22})}{u^2 + \gamma u + 2\omega_0^2}, \quad \bar{\sigma}_{12}(u) = \frac{u\bar{\sigma}_{11}(u)}{2}, \tag{4.151}$$

where

$$g_1(u) = (u^2 + \Gamma u + \Omega_0^2 - \omega_0^2)\left[\Omega_0^2 u + (\gamma + \Gamma + u)(u^2 + \gamma u + \omega_0^2)\right] \tag{4.152}$$

and

$$g_2(u) = \omega_0^2(\gamma + \Gamma + 2u)(u^2 + \gamma u + \omega_0^2 - \Omega_0^2). \tag{4.153}$$

Note that the initial values $x_0$ and $v_0$ are only related to the coefficients $M_1$ and $M_2$, and these functions decay to zero exponentially. The coefficients $\sigma_{ij}$ of the stationary solution given in the previous section can be verified from Eqs. (4.144)-(4.151) by taking $u \to 0$ ($t \to \infty$).

The solution for the PDF is obtained by inverting the Fourier transform

$$\rho\left(\{\mathbf{x}\}, t | \{\mathbf{x_0}\}, 0\right) = \frac{1}{(2\pi)^4} \int e^{i(x_n - M_n)k_n - \frac{\sigma_{nl} k_n k_l}{2}} d^4 k, \quad n = 1, 2, 3, 4. \tag{4.154}$$

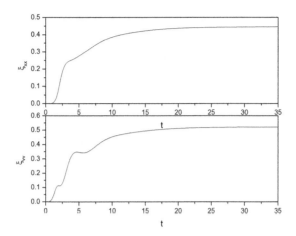

Fig. 4.4 Plots of the variance for the position and velocity ($\langle \zeta_{xx}^2 \rangle$ and $\langle \zeta_{vv}^2 \rangle$) in function of time for $\omega_0 = 2$. The parameters are given by $\epsilon = 1.2$, $\gamma = 0.2$, $\Gamma = 0.2$ and $\Omega_0^2 = 0.7$.

The integration of Eq. (4.154) yields

$$\rho\left(\{\mathbf{x}\}, t| \{\mathbf{x_0}\}, 0\right) = \frac{e^{-\frac{1}{2}\left(\frac{(x-M_1)^2}{\sigma_{11}} + \frac{(v-M_2-(x-M_1)a_{12})^2}{\sigma_{22}-\sigma_{11}a_{12}^2} + \frac{A_3^2}{A_1} + \frac{A_5^2}{A_4}\right)}}{(2\pi)^2 \sqrt{\sigma_{11}\left(\sigma_{22} - \sigma_{11}a_{12}^2\right) A_1 A_4}} \tag{4.155}$$

where

$$A_1 = \sigma_{33} - \sigma_{11}a_{13}^2 - \frac{\left(\sigma_{23} - \sigma_{12}a_{13}\right)^2}{\sigma_{22} - \sigma_{11}a_{12}^2}, \tag{4.156}$$

$$A_2 = \sigma_{34} - \sigma_{11}a_{13}a_{14} - \frac{\left(\sigma_{23} - \sigma_{12}a_{13}\right)\left(\sigma_{24} - \sigma_{12}a_{14}\right)}{\sigma_{22} - \sigma_{11}a_{12}^2}, \tag{4.157}$$

$$A_3 = y - (x - M_1)a_{13} - \frac{(v - M_2 - (x - M_1)a_{12})\left(\sigma_{23} - \sigma_{12}a_{13}\right)}{\sigma_{22} - \sigma_{11}a_{12}^2} \tag{4.158}$$

$$A_4 = \sigma_{44} - \sigma_{11}a_{14}^2 - \frac{\left(\sigma_{24} - \sigma_{12}a_{14}\right)^2}{\sigma_{22} - \sigma_{11}a_{12}^2} - \frac{A_2^2}{A_1}, \tag{4.159}$$

$$A_5 = s - (x - M_1)a_{14} - \frac{(v - M_2 - (x - M_1)a_{14})\left(\sigma_{24} - \sigma_{12}a_{14}\right)}{\sigma_{22} - \sigma_{11}a_{12}^2} - \frac{A_2 A_3}{A_1}. \tag{4.160}$$

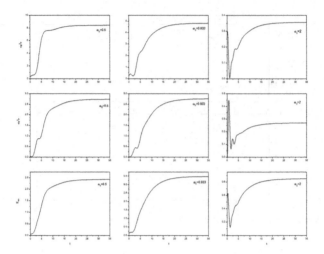

Fig. 4.5    Plots of the second moments for the position and velocity ($\langle x^2 \rangle$ and $\langle v^2 \rangle$), and the mean energy $\langle E_{HO} \rangle$ in function of time for different values of $\omega_0$. The parameters are given by $x_0 = 0.5$, $v_0 = 0.2$, $\epsilon = 1.2$, $\gamma = 1.1$, $\Gamma = 0.2$ and $\Omega_0^2 = 0.7$.

The PDF (4.155) can be numerically checked by substituting it into the Fokker-Planck equation (4.101). To do so, the derivatives of Eq. (4.101) are replaced by the finite differences as follows:

$$\frac{\rho(x, v, y, s, t + h) - \rho(x, v, y, s, t - h)}{2h}$$

$$= -v\frac{\rho(x + h, v, y, s, t) - \rho(x - h, v, y, s, t)}{2h} + (\gamma + \Gamma)\rho(x, v, y, s, t)$$

$$+ \left(\gamma v + \omega_0^2 x - y\right)\frac{\rho(x, v + h, y, s, t) - \rho(x, v - h, y, s, t)}{2h}$$

$$- s\frac{\rho(x, v, y + h, s, t) - \rho(x, v, y - h, s, t)}{2h}$$

$$+ \left(\Gamma s + \Omega_0^2 y\right)\frac{\rho(x, v, y, s + h, t) - \rho(x, v, y, s - h, t)}{2h}$$

$$+ \epsilon\Omega_0^4\frac{\rho(x, v, y, s + h, t) - 2\rho(x, v, y, s, t) + \rho(x, v, y, s - h, t)}{h^2}. \quad (4.161)$$

The first two moments, variance and the mean energies [10] of the driving harmonic noise and linear oscillator can be calculated from the solution (4.155) which are given by

$$\langle x \rangle = M_1(t), \quad (4.162)$$

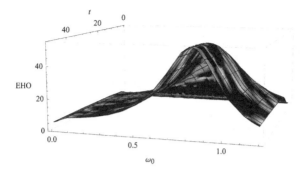

Fig. 4.6  Three-dimensional plot of the mean energy $\langle E_{HO} \rangle$. The initial values and the parameters are given by $x_0 = 0.5$, $v_0 = 0.2$, $\epsilon = 1.2$, $\gamma = 0.2$, $\Gamma = 0.2$ and $\Omega_0^2 = 0.7$.

$$\langle v \rangle = M_2(t), \tag{4.163}$$

$$\langle x^2 \rangle = M_1^2(t) + \sigma_{11}(t), \tag{4.164}$$

$$\langle v^2 \rangle = M_2^2(t) + \sigma_{22}(t), \tag{4.165}$$

$$\zeta_{xx} = \left\langle (x - \langle x \rangle)^2 \right\rangle = \sigma_{11}(t), \tag{4.166}$$

$$\zeta_{vv} = \left\langle (v - \langle v \rangle)^2 \right\rangle = \sigma_{22}(t), \tag{4.167}$$

$$\langle y^2 \rangle = \sigma_{33}(t), \tag{4.168}$$

$$\langle s^2 \rangle = \sigma_{44}(t), \tag{4.169}$$

$$\langle E_{HN} \rangle = \frac{\langle s^2 \rangle}{2} + \Omega_0^2 \frac{\langle y^2 \rangle}{2} = \frac{\sigma_{44}(t) + \Omega_0^2 \sigma_{33}(t)}{2}, \tag{4.170}$$

and

$$\langle E_{HO} \rangle = \frac{\langle v^2 \rangle}{2} + \omega_0^2 \frac{\langle x^2 \rangle}{2} = \frac{M_2^2(t) + \sigma_{22}(t) + \omega_0^2 \left( M_1^2(t) + \sigma_{11}(t) \right)}{2}. \tag{4.171}$$

Note that the mean energies (4.170) and (4.171) reduce to the stationary solutions for $t \to \infty$.

Fig. 4.3 shows the second moments $\langle x^2 \rangle$ and $\langle v^2 \rangle$, and the mean energy $\langle E_{HO} \rangle$ versus time for different values of $\omega_0$. The strong oscillations for $\omega_0 = 2$ are due to the coefficients $M_1$ and $M_2$ which are linked to the initial values $x_0$ and $v_0$. As can be confirmed in Fig. 4.4 for the variances which do not depend on $M_1$ and $M_2$. This means that the oscillations of the system are also strongly dependent on the initial values. In fact, the variance is a measure of the dispersion around the mean ($\langle x \rangle$ and $\langle v \rangle$). It can also suppress the behavior of some forces such as the load force [12]. The resonance occurs at $\omega_0 = 0.824$ and the mean energy is given by $\langle E_{HO} \rangle = 54.21$, and they are in good agreement with the stationary results given in section 4.4.1. In Fig. 4.5 the damping parameter of the oscillator $\gamma$ is changed to $\gamma = 1.1$, and the oscillations are attenuated for $\omega_0 = 2$; the resonance occurs at $\omega_0 = 0.933$ and the mean energy of the oscillator is given by $\langle E_{HO} \rangle = 3.485$, and they are also in good agreement with the stationary results. The results show that not only the parameters $\epsilon$, $\gamma$, $\Gamma$ and $\Omega$ can change the behaviors of the system, but also the initial values $x_0$ and $v_0$. In fact, the strongly oscillatory behaviors shown in Fig. 4.3 are linked to the initial values. Fig. 4.6 shows the evolution of the mean energy $\langle E_{HO} \rangle$ in three-dimensions.

## 4.5   Fokker-Planck equations of the relativistic Brownian motion

The Brownian particles that move randomly through a surrounding medium can be described by the usual nonrelativistic diffusion equation [1–3]. Extension of the classical Brownian motion to the relativistic regime has been proposed by using several approaches [13–18]. In the case of using the Langevin approach two types of Langevin equation have been considered [15–18]. One of them [15] the authors have postulated a constant diffusion coefficient and the corresponding Langevin equation generates only one Fokker-Planck equation. Whereas, the other one [17] the authors have used the relativistic equation of motions and, as a consequence, the relativistic Langevin equation has a multiplicative noise term. In this last case, the one-to-one correspondence between the Fokker-Planck equation and the Langevin equation can not be assured due to the different orders of prescription in stochastic calculus. Consequently, this Langevin equation can generate many different Fokker-Planck equations. For convenience, we consider only the Stratonovich, Ito and Hänggi-Klimontovich or transport prescriptions.

The Fokker-Planck equations and their stationary solutions obtained from the relativistic Brownian motion [17–20], in the laboratory frame, are given by

$$\frac{\partial \rho(\mathbf{p}, t)}{\partial t} = \frac{\partial j_{S/I/HK}^i(\mathbf{p}, t)}{\partial p^i}, \qquad (4.172)$$

where $t$ is the time, $p^i$ are the relativistic momenta, $j_{S/I/HK}^i(\mathbf{p}, t)$ are the probability currents and the indices are the abbreviations of Stratonovich, Ito and Hänggi-Klimontovich or transport prescriptions, respectively. For the one-dimensional case the probability currents are given by

$$j_S = -\left[\nu p \rho(p, t) + D\sqrt{\gamma(p)}\frac{\partial}{\partial p}\left(\sqrt{\gamma(p)}\rho(p, t)\right)\right], \qquad (4.173)$$

$$j_I = -\left[\nu p \rho(p, t) + D\frac{\partial}{\partial p}\left(\gamma(p)\rho(p, t)\right)\right] \qquad (4.174)$$

and

$$j_{HK} = -\left[\nu p \rho(p, t) + D\gamma(p)\frac{\partial}{\partial p}\left(\rho(p, t)\right)\right], \qquad (4.175)$$

where $\gamma(p) = (1 + p^2/(mc)^2)^{1/2}$. For three-dimensional case the probability currents are given by

$$j_S^i = -\left[\nu p^i \rho(p, t) + D(\mathbf{L}^{-1})_k^i\frac{\partial}{\partial p_j}\left(((\mathbf{L}^{-1})^T)_j^k \rho(p, t)\right)\right], \qquad (4.176)$$

$$j_I^i = -\left[\nu p^i \rho(p, t) + D\frac{\partial}{\partial p_j}\left((\mathbf{A}^{-1})_j^i \rho(p, t)\right)\right] \qquad (4.177)$$

and

$$j_{HK}^i = -\left[\nu p^i \rho(p, t) + D(\mathbf{A}^{-1})_j^i\frac{\partial}{\partial p_j}\left(\rho(p, t)\right)\right], \qquad (4.178)$$

where $\mathbf{A}$ and $\mathbf{L}$ are the matrices associated with the momentum components [18].

The stationary solutions of Eq.(4.172) are obtained by setting $\frac{\partial \rho}{\partial t} = 0$ and $j_{S/I/HK}^i = 0$. For the one-dimensional case we have

$$\rho_S(p) = \frac{C_{1S}\exp\left(-\beta\sqrt{1 + \frac{p^2}{m^2c^2}}\right)}{\left(1 + \frac{p^2}{m^2c^2}\right)^{1/4}}, \qquad (4.179)$$

$$\rho_I(p) = \frac{C_{1I}\exp\left(-\beta\sqrt{1 + \frac{p^2}{m^2c^2}}\right)}{\left(1 + \frac{p^2}{m^2c^2}\right)^{1/2}} \qquad (4.180)$$

and

$$\rho_{HK}(p) = C_{1HK} \exp\left(-\beta\sqrt{1 + \frac{p^2}{m^2c^2}}\right), \tag{4.181}$$

where $C_{1S}$, $C_{1I}$ and $C_{1HK}$ are the normalization factors, $\beta = \nu m^2 c^2/D = mc^2/(k_B T)$, $k_B$ is the Boltzmann constant and $T$ is the temperature. It should be noted that only the Hänggi-Klimontovich or transport prescription is consistent with the relativistic Maxwell distribution.

In the three-dimensional case the stationary solutions are given by

$$\rho_S(\mathbf{p}) = \frac{C_{3S} \exp\left(-\beta\sqrt{1 + \frac{\mathbf{p}^2}{m^2c^2}}\right)}{\left(1 + \frac{\mathbf{p}^2}{m^2c^2}\right)^{\frac{3}{4}}}, \tag{4.182}$$

$$\rho_I(\mathbf{p}) = \frac{C_{3I} \exp\left(-\beta\sqrt{1 + \frac{\mathbf{p}^2}{m^2c^2}}\right)}{\left(1 + \frac{\mathbf{p}^2}{m^2c^2}\right)^{\frac{3}{2}}} \tag{4.183}$$

and

$$\rho_{HK}(\mathbf{p}) = C_{3HK} \exp\left(-\beta\sqrt{1 + \frac{\mathbf{p}^2}{m^2c^2}}\right), \tag{4.184}$$

where $C_{3S}$, $C_{3I}$ and $C_{3HK}$ are the normalization factors.

The above stationary solutions can be normalized analytically. The normalization of the stationary solutions is required, such as for the analysis of the second moments in terms of the temperature. For the one-dimensional case the normalized solutions are given by

$$\rho_S(p) = \frac{\sqrt{2\pi} \exp\left(-\beta\sqrt{1 + \frac{p^2}{m^2c^2}}\right)}{2mc\sqrt{\beta}K_{\frac{3}{4}}(\frac{\beta}{2})K_{\frac{1}{4}}(\frac{\beta}{2})\left(1 + \frac{p^2}{m^2c^2}\right)^{1/4}}, \tag{4.185}$$

$$\rho_I(p) = \frac{\exp\left(-\beta\sqrt{1 + \frac{p^2}{m^2c^2}}\right)}{2mcK_0(\beta)\left(1 + \frac{p^2}{m^2c^2}\right)^{1/2}} \tag{4.186}$$

and

$$\rho_{HK}(p) = \frac{\exp\left(-\beta\sqrt{1 + \frac{p^2}{m^2c^2}}\right)}{2mcK_1(\beta)}, \qquad (4.187)$$

where $K_\nu(z)$ denotes the modified Hankel function. The corresponding second moments are given by

$$\left\langle \frac{p^2}{m^2c^2} \right\rangle_S = -\frac{1}{2\sqrt{\beta}K_{\frac{3}{4}}(\frac{\beta}{2})K_{\frac{1}{4}}(\frac{\beta}{2})}$$

$$\times \frac{\partial}{\partial\beta}\left\{\sqrt{\beta}\left[K_{\frac{5}{4}}\left(\frac{\beta}{2}\right)K_{\frac{3}{4}}\left(\frac{\beta}{2}\right) - K_{\frac{1}{4}}\left(\frac{\beta}{2}\right)K_{-\frac{1}{4}}\left(\frac{\beta}{2}\right)\right]\right\}, \qquad (4.188)$$

$$\left\langle \frac{p^2}{m^2c^2} \right\rangle_I = \frac{K_1(\beta)}{\beta K_0(\beta)} \qquad (4.189)$$

and

$$\left\langle \frac{p^2}{m^2c^2} \right\rangle_{HK} = \frac{K_2(\beta)}{\beta K_1(\beta)}. \qquad (4.190)$$

In the case of three-dimensional processes we can also obtain the analytical normalization factors which are given by

$$C_{3S}^{-1} = (mc)^3\sqrt{2\pi\beta}\left[K_{\frac{5}{4}}\left(\frac{\beta}{2}\right)K_{\frac{3}{4}}\left(\frac{\beta}{2}\right) - K_{\frac{1}{4}}\left(\frac{\beta}{2}\right)K_{-\frac{1}{4}}\left(\frac{\beta}{2}\right)\right], \qquad (4.191)$$

$$C_{3I}^{-1} = 2\pi(mc)^3\left\{\beta\pi - {}_2 2F_3\left[-\frac{1}{2}, -\frac{1}{2}; \frac{1}{2}, \frac{1}{2}, 1; \frac{\beta^2}{4}\right]\right.$$

$$\left. + \ln\left(\frac{4}{\beta^2}\right) {}_1F_2\left[-\frac{1}{2}; \frac{1}{2}, 1; \frac{\beta^2}{4}\right] - \sum_{n=0}^{\infty}\frac{2^{1-2n}\beta^{2n}\psi(1+n)}{(-1+2n)\left(\Gamma(1+n)\right)^2}\right\} \qquad (4.192)$$

and

$$C_{3HK}^{-1} = \frac{4\pi(mc)^3 K_2(\beta)}{\beta}, \qquad (4.193)$$

where $\psi(z)$ is the Psi function and ${}_pF_q[a_1, ...., a_p; b_1, ..., b_q; z]$ is the generalized hypergeometric function.

The corresponding second moments of the momentum are given by

$$\left\langle \frac{\mathbf{p}^2}{m^2 c^2} \right\rangle_S = C_{3S} \frac{\partial^2}{\partial \beta^2} \left( \frac{1}{C_{3S}} \right) - 1, \tag{4.194}$$

$$\left\langle \frac{\mathbf{p}^2}{m^2 c^2} \right\rangle_I = \frac{4\pi (mc)^3 C_{3I} K_1(\beta)}{\beta} - 1, \tag{4.195}$$

and

$$\left\langle \frac{\mathbf{p}^2}{m^2 c^2} \right\rangle_{HK} = \frac{1}{2} \left[ \frac{K_4(\beta)}{K_2(\beta)} - 1 \right]. \tag{4.196}$$

## 4.6   Appendices

### 4.6.1   *Numerical inversion of Laplace transforms*

A useful formula for the numerical inversion of Laplace transforms was obtained by H. Stehfest [11]. It is given by

$$f(t) = \frac{\ln(2)}{t} \sum_{i=1}^{n/2} \left\{ \frac{(-1)^{i-1} i}{\left(\frac{n}{2}\right)!} \binom{\frac{n}{2}}{i} \left( \frac{n}{2} - i + 1 \right)^{\frac{n}{2}-1} \frac{\left( 2 \left( \frac{n}{2} - i + 1 \right) \right)!}{\left( \frac{n}{2} - i + 1 \right)! \left( \frac{n}{2} - i \right)!} \right.$$

$$\left. \times \sum_{j=0}^{n/2-i+1} \left[ (-1)^j \binom{\frac{n}{2} - i + 1}{j} f_s \left( \left( \frac{n}{2} - i + j + 1 \right) \frac{\ln(2)}{t} \right) \right] \right\}, \tag{4.197}$$

where $f_s(s)$ is the Laplace transform of $f(t)$; $n$ is an integer number and it must be even. In general, the numerical inversion formula (4.197) gives good precision, and the number $n$ given for computation is usually less than 24.

## Bibliography

[1]  H. Risken, *The Fokker-Planck Equation*, 2th ed. (Springer-Verlag, Berlin, 1996).

[2]  W. T. Coffey, Y. P. Kalmykov and J. T. Waldron, *The Langevin equation* (World Scientific, Singapore, 2005).

[3]  S. Chandrasekhar, *Rev. Mod. Phys.* **15**, 1 (1943).

[4]  M. C. Wang and G. E. Uhlenbeck, *Rev. Mod. Phys.* **17**, 323 (1945).

[5] W. Kaplan, *Advanced Calculus*, 4th ed. (Addison-Wesley, USA, 1993).

[6] F. Moss, P. V. E. McClintock (eds.), *Noise in nonlinear dynamical systems*, Vol. 1-3 (Cambridge University Press, Cambridge, 1989).

[7] L. Schimansky-Geier and Ch. Ziilicke, *Z. Phys. B* **79**, 451 (1990).

[8] J. J. Hesse and L. Schimansky-Geier, *Z. Phys. B* **84**, 467 (1991).

[9] A. Neiman and L. Schimansky-Geier, *Phys. Rev. Lett.* **72**, 2988 (1994).

[10] J. Masoliver and J. M. Porrà, *Phys. Rev. E* **48**, 4309 (1993).

[11] H. Stehfest, *Commun. ACM* **13**, 47 (1970).

[12] K. S. Fa, *Eur. J. Phys.* **37**, 065101 (2016).

[13] R. Hakim, *J. Math. Phys.* **6**, 1482 (1965).

[14] U. Ben-Yaacov, *Phys. Rev. D* **23**, 1441 (1981).

[15] F. Debbasch, K. Mallick and J. P. Rivet, *J. Stat. Phys.* **88**, 945 (1997).

[16] F. Debbasch and J. P. Rivet, *J. Stat. Phys.* **90**, 1179 (1998).

[17] J. Dunkel and P. Hänggi, *Phys. Rev. E* **71**, 016124 (2005).

[18] J. Dunkel and P. Hänggi, *Phys. Rev. E* **72**, 036106 (2005).

[19] J. Dunkel and P. Hänggi, *Phys. Rev. E* **74**, 051106 (2006).

[20] J. Dunkel and P. Hänggi, *Phys. Rep.* **471**, 1 (2009).

# Chapter 5

# Generalized Langevin equations

## 5.1 Introduction

In this chapter we will consider some generalized Langevin equations which contain non-local operators in time. One of the most interesting features incorporated into the Langevin equation is that related to the memory effect. In particular, the memory effect incorporated into the Langevin approach can be associated with the retardation of friction and fractal media [1–7]. Moreover, it has been suggested to substitute the ordinary derivative by a fractional derivative when separation of the microscopic and macroscopic time scales does not exist [2,3]. The first case is the generalization of the friction term in the Langevin equation. This is required for describing more general types of Brownian motion [1, 4–13]. We may say that if the noise and dissipation are associated with the same source, then the noise is internal. If they are not associated with the same source, then the noise is external. For internal colored noise the friction term assumes a more general form which contains a non-local operator with memory, and it is described by

$$m\frac{dv}{dt} + m\int_0^t dt_1\gamma\left(t - t_1\right)v(t_1) = F(x) + Q(t),\qquad(5.1)$$

where $v$ is the velocity, $\gamma(t)$ is the dissipative memory kernel, $F(x)$ is an external force and $Q(t)$ is assumed to be a Gaussian random force with zero mean $\langle Q(t)\rangle = 0$ and correlation function given by

$$\langle Q(t_1)Q(t_2)\rangle = C\left(|t_1 - t_2|\right).\qquad(5.2)$$

Note that if the dissipative memory kernel is given by $\gamma(t) = 1/[t^\alpha\Gamma(1-\alpha)]$, then Eq. (5.1) becomes a fractional Langevin equation given by

$$m\frac{d^2x}{dt^2} + m\,_0^C D_t^\alpha x = F(x) + Q(t),\qquad(5.3)$$

where $_0^C D_t^\alpha x$ is the Caputo fractional derivative [14] defined by

$$_0^C D_t^\alpha x = \frac{1}{\Gamma(1-\alpha)} \int_0^t dt_1 \frac{\frac{dx}{dt_1}}{(t-t_1)^\alpha}, \quad 0 < \alpha < 1. \tag{5.4}$$

For a system which is in equilibrium state, the internal noise is related to the dissipative memory kernel as follows:

$$m\gamma(t) = \frac{C(t)}{k_B T}. \tag{5.5}$$

The relation (5.5) is called the second fluctuation-dissipation theorem. The dissipative memory kernel $\gamma(t_2 - t_1)$ depends only on the time difference means that the bath is in a stationary distribution [15]. Besides, the colored power spectral density of the thermal noise, which is the Fourier transform of Eq. (5.5), has been confirmed by experiments recently [16].

## 5.2    Derivation of the generalized Langevin equation from the Hamiltonian formalism

The generalized Langevin equation (5.1) can be derived from the Hamiltonian formalism [1, 8, 12, 13]. Let us consider the phase space of the total system described by the Hamiltonian

$$H = H(q_1, q_2, ...., q_n, p_1, p_2, ....., p_n, t) \tag{5.6}$$

where $q_i$ and $p_i$ denote the generalized coordinates and generalized momenta of the system. The equation of motion is described by [17, 18]

$$\frac{\partial q_j}{\partial t} = \frac{\partial H}{\partial p_j}, \tag{5.7}$$

and

$$\frac{\partial p_j}{\partial t} = -\frac{\partial H}{\partial q_j}. \tag{5.8}$$

Now one considers a particle in one dimension that is in contact with a thermal reservoir (heat bath) which is formed by a collection of harmonic oscillators. For a bilinear coupling to the system, between the coordinates $q$ and $x_\alpha$ (bath), the corresponding Hamiltonian is described by

$$H = \frac{p^2}{2m} + V(q) + \sum_\alpha \left[ \frac{p_\alpha^2}{2m_\alpha} + \frac{m_\alpha \omega_\alpha^2}{2} \left( x_\alpha + \frac{c_\alpha}{m_\alpha \omega_\alpha^2} q \right)^2 \right], \tag{5.9}$$

where the index $\alpha$ runs over all the bath degrees of freedom, $\omega_\alpha$ are the harmonic bath frequencies, $m_\alpha$ are the harmonic bath masses, $x_\alpha$ are the coordinates related to the bath and $c_\alpha$ are the coupling constants between the bath and the coordinate $q$. The coordinate $q$ is assumed to be subject to a potential $V(q)$ as well. Note that for deriving the generalized Langevin equation (5.1) we need to derive from the Hamiltonian the detailed dynamics of $q$ alone. Now substituting the Hamiltonian (5.9) into Eqs. (5.7) and (5.8) yields

$$\frac{dq}{dt} = \frac{p}{m}, \tag{5.10}$$

$$\frac{dp}{dt} = -\frac{dV(q)}{dq} - \sum_\alpha \left[ c_\alpha x_\alpha + \frac{c_\alpha^2}{m_\alpha \omega_\alpha^2} q \right], \tag{5.11}$$

$$\frac{dx_\alpha}{dt} = \frac{p_\alpha}{m_\alpha}, \tag{5.12}$$

and

$$\frac{dp_\alpha}{dt} = -m_\alpha \omega_\alpha^2 x_\alpha - c_\alpha q. \tag{5.13}$$

To solve the set of equations (5.10)-(5.13) we first eliminate the momenta $p$ and $p_\alpha$, and the result is given by

$$m\frac{d^2 q}{dt^2} = -\frac{dV(q)}{dq} - \sum_\alpha \left[ c_\alpha x_\alpha + \frac{c_\alpha^2}{m_\alpha \omega_\alpha^2} q \right] \tag{5.14}$$

and

$$m_\alpha \frac{d^2 x_\alpha}{dt^2} = -m_\alpha \omega_\alpha^2 x_\alpha - c_\alpha q. \tag{5.15}$$

Now we solve Eq. (5.15) to obtain a solution for $x_\alpha$. Applying the Laplace transform to Eq. (5.15) yields

$$x_{s\alpha}(s) = \frac{x_\alpha(0)s}{\omega_\alpha^2 + s^2} + \frac{\dot{x}_\alpha(0)}{\omega_\alpha^2 + s^2} - \frac{c_\alpha \bar{q}}{\omega_\alpha^2 + s^2}, \tag{5.16}$$

where the dot denotes the derivative with respect to time. The solution for $x_\alpha$ is obtained by applying the inverse Laplace transform and convolution theorem to Eq. (5.16), and it is given by

$$x_\alpha(t) = x_\alpha(0) \cos(\omega_\alpha t) + \frac{\dot{x}_\alpha(0) \sin(\omega_\alpha t)}{\omega_\alpha} -$$

$$\frac{c_\alpha}{m_\alpha \omega_\alpha} \int_0^t d\tau q(\tau) \sin(\omega_\alpha(t - \tau)). \tag{5.17}$$

Integrating by parts the third term of the right-hand side of Eq. (5.17) we arrive at

$$x_\alpha(t) = x_\alpha(0)\cos(\omega_\alpha t) + \frac{\dot{x}_\alpha(0)\sin(\omega_\alpha t)}{\omega_\alpha}$$

$$+ \frac{c_\alpha}{m_\alpha \omega_\alpha^2} \int_0^t d\tau \dot{q}(\tau)\cos(\omega_\alpha(t-\tau)) - \frac{c_\alpha}{m_\alpha \omega_\alpha^2}[q(t) - q(0)\cos(\omega_\alpha t)]. \quad (5.18)$$

Substituting Eq. (5.18) into (5.14) we obtain

$$m\ddot{q} = -\frac{dV(q)}{dq} - \sum_\alpha \frac{c_\alpha^2}{m_\alpha \omega_\alpha^2} \int_0^t d\tau \dot{q}(\tau)\cos(\omega_\alpha(t-\tau))$$

$$- \sum_\alpha c_\alpha \left[ x_\alpha(0)\cos(\omega_\alpha t) + \frac{\dot{x}_\alpha(0)\sin(\omega_\alpha t)}{\omega_\alpha} + \frac{c_\alpha}{m_\alpha \omega_\alpha^2} q(0)\cos(\omega_\alpha t) \right]. \quad (5.19)$$

Eq. (5.19) can be rewritten as follows:

$$m\ddot{q} = -\frac{dV(q)}{dq} - \int_0^t d\tau \dot{q}(\tau)\bar{\gamma}(t-\tau) + Q(t), \quad (5.20)$$

where

$$\bar{\gamma}(t) = \sum_\alpha \frac{c_\alpha^2}{m_\alpha \omega_\alpha^2}\cos(\omega_\alpha(t)) \quad (5.21)$$

and

$$Q(t) = -\sum_\alpha c_\alpha$$

$$\times \left[ \left( x_\alpha(0) + \frac{c_\alpha}{m_\alpha \omega_\alpha^2} q(0) \right)\cos(\omega_\alpha t) + \frac{p_\alpha(0)\sin(\omega_\alpha t)}{m_\alpha \omega_\alpha} \right]. \quad (5.22)$$

Thus, we have obtained the generalized Langevin equation described in Eq. (5.1).

### 5.2.1   *General analysis: First two moments, variances, co-variance, PDF and Fokker-Planck equation*

Note that general expressions for the first two moments, variance and covariance can be determined from Eq. (5.1), in the case of linear force. Moreover, a solution of Eq. (5.1) can also be obtained. For convenience we consider a unit mass ($m = 1$) hereafter. For obtaining the first two moments, variance and covariance we use the Laplace transform technique. We consider $F(x) = -\left(\omega^2 x - F_L\right)$; the first term on the right-hand side is the force of a harmonic potential which is a confining potential, whereas the second one is a constant load force which conducts the particles of the system to its direction. The initial conditions are expressed by $x_0 = x(0)$ and $v_0 = v(0)$. Applying the Laplace transform and the convolution theorem to Eq. (5.1) yields

$$s^2 x_s(s) - sx_0 - v_0 + \gamma_s \left(sx_s(s) - sx_0\right) + \omega^2 x_s(s) - \frac{F_L}{s} = Q_s(s). \quad (5.23)$$

We rewrite Eq. (5.23) as follows:

$$x_s(s) = \frac{x_0 \left(s + \gamma_s\right) + v_0 + \frac{F_L}{s} + Q_s}{\omega^2 + s\gamma_s + s^2}$$

$$= x_0 \left(\frac{1}{s} - \frac{\omega^2 G_s}{s}\right) + v_0 G_s + F_L \frac{G_s}{s} + Q_s G_s, \quad (5.24)$$

where

$$G_s(s) = \frac{1}{\omega^2 + s\gamma_s + s^2}. \quad (5.25)$$

Applying the inverse Laplace transform and the convolution theorem to Eq. (5.24) we obtain

$$x(t) = \langle x \rangle + \int_0^t dt_1 G\left(t - t_1\right) Q(t_1), \quad (5.26)$$

where

$$\langle x \rangle = x_0 \left[1 - \omega^2 I\left(t\right)\right] + v_0 G\left(t\right) + F_L I(t) \quad (5.27)$$

and

$$I(t) = \int_0^t dt_1 G(t_1). \quad (5.28)$$

The kernel $G\left(t\right)$ is calculated through the Laplace inversion of $G_s$. Note that Eq. (5.26) is a linear transformation of $Q(t)$. Thus for a Gaussian process $Q(t)$, any linear transformation of it $x(t)$ is also a Gaussian [1]. This consideration is important for obtaining the PDF and the corresponding Fokker-Planck equation of the generalized Langevin equation in the presence of harmonic potential.

From Eq. (5.26) we can obtain the velocity $v(t)$ which is given by

$$v(t) = \langle v \rangle + \int_0^t dt_1 g\left(t - t_1\right) Q(t_1) \quad (5.29)$$

with $G(0) = 0$, where

$$\langle v \rangle = \left(F_L - x_0\omega^2\right)G(t) + v_0 g(t) \tag{5.30}$$

and

$$g(t) = \frac{dG(t)}{dt}. \tag{5.31}$$

The variance and covariance are obtained from the solutions (5.26) and (5.29). First, we rewrite Eqs. (5.26) and (5.29) as follows:

$$x(t) = \langle x \rangle + \int_0^t dt_1 G(t_1) Q(t - t_1) \tag{5.32}$$

and

$$v(t) = \langle v \rangle + \int_0^t dt_1 g(t_1) Q(t - t_1). \tag{5.33}$$

From Eq. (5.32) we have

$$\sigma_{xx}(t) = \left\langle (x - \langle x \rangle)^2 \right\rangle$$

$$= \left\langle \int_0^t dt_1 G(t_1) \int_0^t dt_2 G(t_2) Q(t - t_1)Q(t - t_2) \right\rangle. \tag{5.34}$$

Using the relation (5.2) we obtain

$$\sigma_{xx}(t) = \int_0^t dt_1 G(t_1) \int_0^t dt_2 G(t_2) C\left(|t_1 - t_2|\right). \tag{5.35}$$

Now we divide the area of integration into two areas separated at $t_1 = t_2$, and Eq. (5.35) can be written as follows.

$$\sigma_{xx}(t) = \int_0^t dt_1 G(t_1) \int_0^{t_1} dt_2 G(t_2) C(t_1 - t_2)$$

$$+ \int_0^t dt_2 G(t_2) \int_0^{t_2} dt_1 G(t_1) C(t_2 - t_1), \tag{5.36}$$

where the first and second terms correspond to $t_1 > t_2$ and $t_2 > t_1$, respectively. Changing the variables $t_1 \leftrightarrow t_2$ in the second term of the right-hand side of Eq. (5.36) we arrive at

$$\sigma_{xx}(t) = 2 \int_0^t dt_1 G(t_1) \int_0^{t_1} dt_2 G(t_2) C(t_1 - t_2). \tag{5.37}$$

For $\sigma_{xv}(t)$ we have

$$\sigma_{xv}(t) = \langle (x - \langle x \rangle)(v - \langle v \rangle) \rangle$$

$$= \left\langle (x - \langle x \rangle) \frac{d}{dt}(x - \langle x \rangle) \right\rangle = \frac{1}{2}\frac{d}{dt}\left\langle (x - \langle x \rangle)^2 \right\rangle = \frac{1}{2}\frac{d\sigma_{xx}}{dt}. \tag{5.38}$$

Substituting Eq. (5.37) into Eq. (5.38) yields

$$\sigma_{xv}(t) = G(t) \int_0^t dt_1 G(t_1) C(t - t_1). \tag{5.39}$$

For $\sigma_{vv}(t)$ we obtain

$$\sigma_{vv}(t) = \langle v^2 \rangle - \langle v \rangle^2 = 2 \int_0^t dt_1 g(t_1) \int_0^{t_1} dt_2 g(t_2) C(t_1 - t_2). \tag{5.40}$$

For the second moments we have

$$\langle x^2 \rangle = \langle x \rangle^2 + \sigma_{xx}(t) \tag{5.41}$$

and

$$\langle v^2 \rangle = \langle v \rangle^2 + \sigma_{vv}(t). \tag{5.42}$$

We can see that the relaxation function $G(t)$ does not contain any influence of the load force. This means that the load force does not affect the behaviors of variances and covariance as can be seen in Eqs. (5.37)-(5.40). However, the load force is important for the first two moments. Let us consider $v_0 = 0$ and $x_0$ placed at a position different from the minimum of potential, i.e., $x_0 \neq 0$. In this case, the load force can be used to cancel the influence of linear force by setting $F_L = x_0\omega^2$, and the particle remains in average at the position $x_0$ with zero velocity independently of the type of dissipative memory kernel.

In the case of $C(t)$ proportional to the dissipative memory kernel $C(t) = c\gamma(t)$ the expressions (5.37)-(5.40) can be simplified. From Eq. (5.25) we have

$$G_s(s)\left(\omega^2 + s\gamma_s + s^2\right) = 1; \tag{5.43}$$

and we obtain

$$G_s(s)\gamma_s(s) = \frac{1}{s} - \omega^2\frac{G_s(s)}{s} - sG_s(s). \tag{5.44}$$

Inverting the Laplace transform yields

$$\int_0^t dt_1 G(t_1)\gamma(t - t_1) = 1 - \omega^2 I - g(t). \tag{5.45}$$

Substituting Eq. (5.45) into Eq. (5.37) we obtain

$$\sigma_{xx}(t) = c \left[ 2I(t) - G^2(t) - \omega^2 I^2(t) \right], \tag{5.46}$$

where the last term has been obtained by using the relation

$$\int_0^t dt_1 \int_0^t dt_2 G(t_1) G(t_2) = 2 \int_0^t dt_1 \int_0^{t_1} dt_2 G(t_1) G(t_2). \tag{5.47}$$

The relation (5.47) is obtained by dividing the integration area into two areas separated at $t_1 = t_2$.

For $\sigma_{xv}(t)$ yields

$$\sigma_{xv}(t) = \frac{1}{2} \frac{d\sigma_{xx}}{dt} = cG(t) \left[ 1 - g(t) - \omega^2 I(t) \right]. \tag{5.48}$$

For $\sigma_{vv}(t)$ we derive the relation (5.45), and we obtain

$$\int_0^t dt_1 g(t_1) \gamma(t - t_1) = -\omega^2 G(t) - \frac{dg(t)}{dt}. \tag{5.49}$$

Substituting Eq. (5.49) into Eq. (5.40) yields

$$\sigma_{vv}(t) = c \left[ 1 - g^2(t) - \omega^2 G^2(t) \right]. \tag{5.50}$$

Eqs. (5.46), (5.48) and (5.50) are appropriate for a system which is in equilibrium state, i.e., when the internal noise is related to the dissipative memory kernel $\gamma(t) = C(t)/k_B T$, where $k_B$ is the Boltzmann constant and $T$ is the absolute temperature of environment.

Further, for the cross correlation $< v(t + \tau)x(t) >$ we have

$$\langle v(t + \tau) x(t) \rangle = \langle v(t + \tau) \rangle \langle x(t) \rangle$$

$$+ \int_0^{t+\tau} dt_1 g(t + \tau - t_1) \int_0^t dt_2 G(t - t_2) C(|t_1 - t_2|). \tag{5.51}$$

This quantity has the process sampled at time $t$ for the position and at time $t + \tau$ for the velocity and then calculated for the ensemble. It measures the correlation between the position and velocity of the system. We see that, from Eq. (5.51), the cross correlation is different from zero and it strongly depends on the initial conditions and parameters of the system. Moreover, for $\tau = 0$ and $C(t) = c\gamma(t)$ we recover the covariance $\sigma_{xv}$ (5.48).

In order to investigate some detail of the above system we consider the exponential correlation function and without external force $F(x) = 0$. We first consider the case of frictional memory kernel given by $\gamma(t) = \gamma_0 e^{-\lambda t}$. In this case the Laplace transform of $\gamma(t)$ is given by $\gamma_s(s) = \gamma_0 / (s + \lambda)$. From (5.25) with $\omega = 0$, we obtain

$$G(t) = \frac{\lambda}{\gamma_0} \left[ 1 - A_1 e^{-\lambda t/2} \sin(\lambda_1 t + \phi) \right], \quad \gamma_0 > \frac{\lambda^2}{4}, \tag{5.52}$$

$$G(t) = \frac{4}{\lambda} \left[ 1 - e^{-\lambda t/2} \left( 1 + \frac{\lambda}{4}t \right) \right] , \qquad \gamma_0 = \frac{\lambda^2}{4} , \qquad (5.53)$$

$$G(t) = \frac{\lambda}{\gamma_0}$$

$$\times \left\{ 1 - \frac{A_2 + 1}{2} e^{-\left(\frac{\lambda}{2} - \lambda_2\right)t} + \frac{A_2 - 1}{2} e^{-\left(\frac{\lambda}{2} + \lambda_2\right)t} \right\} , \quad \gamma_0 < \frac{\lambda^2}{4} , \qquad (5.54)$$

where

$$A_1 - \frac{\gamma_0}{\lambda \lambda_1} , \ A_2 = \frac{\lambda^2 - 2\gamma_0}{2\lambda \lambda_2} , \ \lambda_1 = \sqrt{\gamma_0 - \frac{\lambda^2}{4}} , \ \lambda_2 - \sqrt{\frac{\lambda^2}{4} - \gamma_0} \quad (5.55)$$

and

$$\phi = \arctan \left[ \frac{\lambda \lambda_1}{\frac{\lambda^2}{2} - \gamma_0} \right] . \qquad (5.56)$$

We see that the behavior of $G(t)$ changes with the damping parameters $\gamma_0$ and $\lambda$. In the case of internal noise, the correlation function is related to the dissipative memory kernel given by $C(t) = (D/2\tau_c)e^{-t/\tau_c}$ and $\gamma(t) = C(t)/k_B T$ [6], where $\tau_c$ is the correlation time. Then, $\lambda = 1/\tau_c$ and $\gamma_0 = D/(2\tau_c k_B T)$. For this case, the change of behavior of the kernel $G(t)$ can be associated with the noise intensity $D$. For $D > k_B T/2\tau_c$, $G(t)$ oscillates. For $D = k_B T/2\tau_c$, $G(t)$ does not oscillate, however, it contains a term of type $e^{-z}z$. For $D < k_B T/2\tau_c$, $G(t)$ contains a combination of exponential terms. These behaviors are also obtained for the variances as they are shown below.

From the solutions (5.52)-(5.54) one can obtain the variances $\sigma_{xx}$ and $\sigma_{vv}$ which are given by

$$\sigma_{xx}(t) = -\frac{4\left(k_B T\right)^3 A_1}{D^2}$$

$$\times \left\{ \sin\phi + 2\lambda_1 \tau_c \cos\phi - e^{-t/2\tau_c} \left[ 3\sin\left(\lambda_1 t + \phi\right) + 2\lambda_1 \tau_c \cos\left(\lambda_1 t + \phi\right) \right] \right\}$$

$$+ \frac{\left(2k_B T\right)^2}{D} t - \frac{4\left(k_B T\right)^3}{D^2} \left[ 1 + A^2 e^{-t/\tau_c} \sin^2\left(\lambda_1 t + \phi\right) \right] , \quad D > \frac{k_B T}{2\tau_c}, \quad (5.57)$$

$$\sigma_{xx}(t) = \frac{(2k_BT)^2}{D}\left[t - 4\tau_c + \frac{2D\tau_c^2}{k_BT} - \frac{k_BT}{D}\right]$$

$$+\frac{8\,(k_BT)^2\,\tau_c e^{-t/2\tau_c}}{D}\left[1 + \left(1 + \frac{t}{2\tau_c}\right)\left(1 - \frac{D\tau_c}{k_BT}\right)\right] + \frac{8\,(k_BT)^3\,e^{-t/2\tau_c}}{D^2}$$

$$\times\left[1 + \left(1 - \frac{D\tau_c}{k_BT}\right)\frac{t}{2\tau_c}\right]\left[1 - \frac{e^{-t/2\tau_c}}{2}\right], \quad D = \frac{k_BT}{2\tau_c}, \qquad (5.58)$$

$$\sigma_{xx}(t) = \frac{2\,(k_BT)^2}{D}$$

$$\times\left[2t + \frac{1 + A_2}{\frac{\lambda}{2} - \lambda_2}\left(e^{-(\lambda/2-\lambda_2)t} - 1\right) + \frac{1 - A_2}{\frac{\lambda}{2} + \lambda_2}\left(e^{-(\lambda/2+\lambda_2)t} - 1\right)\right] - \frac{4\,(k_BT)^3}{D^2}$$

$$\times\left[1 - \frac{1 + A_2}{2}e^{-(\lambda/2-\lambda_2)t} - \frac{1 - A_2}{2}e^{-(\lambda/2+\lambda_2)t}\right]^2, \quad D < \frac{k_BT}{2\tau_c}, \quad (5.59)$$

$$\sigma_{vv}(t) = k_BT\left\{1 - \frac{(2k_BT)^2A_1^2}{D^2}e^{-t/\tau_c}\right.$$

$$\left.\times\left[\frac{1}{2\tau_c}\sin(\lambda_1 t + \phi) - \lambda_1\cos(\lambda_1 t + \phi)\right]^2\right\}, \quad D > \frac{k_BT}{2\tau_c}, \qquad (5.60)$$

$$\sigma_{vv}(t) = k_BT\left[1 - \left(1 + \frac{t}{2\tau_c}\right)^2 e^{-t/\tau_c}\right], \quad D = \frac{k_BT}{2\tau_c} \qquad (5.61)$$

and

$$\sigma_{vv}(t) = k_BT - \frac{(2k_BT)^3}{2D^2}\left[\frac{1 + A_2}{2}\left(\frac{\lambda}{2} - \lambda_2\right)e^{-(\lambda/2-\lambda_2)t}\right.$$

$$\left.+ \frac{1 - A_2}{2}\left(\frac{\lambda}{2} + \lambda_2\right)e^{-(\lambda/2+\lambda_2)t}\right]^2, \quad D < \frac{k_BT}{2\tau_c}. \qquad (5.62)$$

It is easy to see that the asymptotic behaviors of $\sigma_{xx}(t)$ are similar and they present normal diffusion which are given by $\sigma_{xx}(t) \sim \frac{(2k_BT)^2}{D}t$, whereas $\sigma_{vv}(t) \sim k_B T$. The former result shows that the internal exponential correlation function does not generate anomalous diffusion processes, in contrast to the power-law correlation function [5, 6].

Note that we have limited the stochastic force $Q(t)$ to a Gaussian distribution due to the linear transformation of $x(t)$ for linear force. In this case, any linear transformation of $x(t)$ is also a Gaussian. This consideration can be used for obtaining the PDF and the corresponding Fokker-Planck equation of the generalized Langevin equation. To do so, we use the characteristic function and the above results for the first moments, variances and covariances. Note that the characteristic function is the Fourier transform of the joint probability density function, then the probability density function is the inverse Fourier transform of the characteristic function. The characteristic function for two random variables is given by [19]

$$C_2(u_1, u_2) = \int e^{i(u_1 x_1 + u_2 x_2)} \rho(x_1, x_2) \, dx_1 dx_2. \qquad (5.63)$$

The joint probability density $\rho(x_1, x_2)$ is obtained by inverting the Fourier transform (5.63) and we have

$$\rho(x_1, x_2) = \frac{1}{(2\pi)^2} \int e^{-i(u_1 x_1 + u_2 x_2)} C_2(u_1, u_2) \, du_1 du_2. \qquad (5.64)$$

Moreover, the characteristic function may be expressed by the moments $M_{n_1, n_2}$

$$C_2(u_1, u_2) = e^{\sum_{n_1, n_2} M_{n_1, n_2} \frac{(iu_1)^{n_1}}{n_1!} \frac{(iu_2)^{n_2}}{n_2!}}, \qquad (5.65)$$

where

$$M_{n_1, n_2} = \left(\frac{\partial}{\partial iu_1}\right)^{n_1} \left(\frac{\partial}{\partial iu_2}\right)^{n_2} C_2(u_1, u_2)|_{u_1 = u_2 = 0}. \qquad (5.66)$$

From the first moments, variances and covariances we obtain the following characteristic function [5, 6]:

$$C_2(u_1, u_2, t) = e^{iu_1 \langle x \rangle + iu_2 \langle v \rangle - \frac{1}{2}\left(u_1^2 \sigma_{xx} + 2u_1 u_2 \sigma_{xv} + u_2^2 \sigma_{vv}\right)}, \qquad (5.67)$$

The corresponding Fokker-Planck equation of the generalized Langevin equation for linear force with $F_L = 0$ can be obtained from the characteristic function (5.67) [5, 6], and it is given by

$$\left[\frac{\partial}{\partial t} + v\frac{\partial}{\partial x} - \bar{\omega}^2(t)x\frac{\partial}{\partial v}\right] \rho(x, v, t)$$

$$= \bar{\gamma}(t)\frac{\partial}{\partial v}\left(v\rho(x, v, t)\right) + \phi(t)\frac{\partial^2 \rho(x, v, t)}{\partial v^2} + \psi(t)\frac{\partial^2 \rho(x, v, t)}{\partial v \partial x}, \qquad (5.68)$$

where

$$\bar{\gamma}(t) = -\frac{d}{dt} \ln \Delta(t), \tag{5.69}$$

$$\bar{\omega}^2(t) = \frac{-G(t)\dot{g}(t) + g^2(t)}{\Delta(t)}, \tag{5.70}$$

$$\Delta(t) = \frac{g(t)}{\omega^2} \left(1 - \omega^2 I(t)\right) + G^2(t), \tag{5.71}$$

$$\phi(t) = \bar{\omega}^2(t)\sigma_{xv}(t) + \bar{\gamma}(t)\sigma_{vv}(t) + \frac{\dot{\sigma}_{vv}(t)}{2} \tag{5.72}$$

and

$$\psi(t) = \dot{\sigma}_{xv}(t) + \bar{\gamma}(t)\sigma_{xv}(t) + \bar{\omega}^2(t)\sigma_{xx}(t) - \sigma_{vv}(t). \tag{5.73}$$

As we have mentioned above the internal noise is related to the dissipative memory kernel when the noise and dissipation are associated to the same source. If the noise and dissipation are not associated to the same source, then the noise is said to be external. Several types of correlation function have been used to study the generalized Langevin equation, for instance, exponential function, power-law function, generalized Mittag-Leffler function and combination of two terms for the dissipative force have been suggested [10, 20–24].

## 5.3   Fractional Langevin equation

A simple generalization of the generalized Langevin Eq. (5.1) is to replace the ordinary derivative by a fractional derivative. Let us consider the following generalized Langevin equation:

$$_0^C D_t^\alpha v + \int_0^t dt_1 \gamma\left(t - t_1\right) v = F(x) + Q(t), \quad \text{for } 0 < \alpha < 1, \tag{5.74}$$

with the same Gaussian random force with mean zero and correlation function given by (5.2). The operator $_0^C D_t^\alpha$ is the Caputo fractional derivative given in Eq. (5.4). We note that the fractional Langevin equation (5.74) with the use of the Riemann-Liouville fractional derivative and white noise has been investigated in [25]. Further, the ordinary Langevin equation with a correlated noise has been investigated in [26] and it can also produce the anomalous diffusion processes.

In the case of Eq. (5.74) with linear external force, we will show that formal expressions for the first moments and general expressions for the variance and covariance can also be obtained. Eq. (5.74) can be solved by using the Laplace transform, with the initial conditions $x_0 = x(0)$ and $v_0 = v(0)$. We also consider a linear force given by $F(x) = -\left(\omega^2 x - F_L\right)$. Applying the Laplace transform and the convolution theorem to Eq. (5.74) yields

$$\frac{s^2 x_s(s) - s x_0 - v_0}{s^{1-\alpha}} + \gamma_s \left(s x_s(s) - x_0\right) = -\omega^2 x_s(s) + \frac{F_L}{s} + Q_s(s). \quad (5.75)$$

We rewrite Eq. (5.75) as follows:

$$x_s(s) = \frac{x_0 \left(s^\alpha + \gamma_s\right) + v_0 s^{\alpha-1} + \frac{F_L}{s} + Q_s}{\omega^2 + s\gamma_0 + s^{1+\alpha}}$$

$$= x_0 \left(\frac{1}{s} - \frac{\omega^2 G_{s\alpha}}{s}\right) + v_0 \frac{G_{s\alpha}}{s^{1-\alpha}} + F_L \frac{G_{s\alpha}}{s} + Q_s G_{s\alpha}, \quad (5.76)$$

where

$$G_{s\alpha}(s) = \frac{1}{\omega^2 + s\gamma_s + s^{1+\alpha}}. \quad (5.77)$$

Applying the inverse Laplace transform and the convolution theorem to Eq. (5.76) we obtain

$$x(t) = \langle x \rangle + \int_0^t dt_1 G_\alpha \left(t - t_1\right) Q(t_1), \quad (5.78)$$

where

$$\langle x \rangle = x_0 \left(1 - \omega^2 I_\alpha(t)\right) + F_L I_\alpha(t) + \frac{v_0}{\Gamma\left(1 - \alpha\right)} \int_0^t dt_1 \frac{G_\alpha\left(t_1\right)}{\left(t - t_1\right)^\alpha} \quad (5.79)$$

and $I_\alpha(t) = \int_0^t dt_1 G_\alpha \left(t - t_1\right)$. The kernel $G_\alpha\left(t\right)$ is obtained from the Laplace inversion of Eq. (5.77). Note that Eq. (5.78) also presents a linear transformation of $Q(t)$. Thus the procedure described in the previous section for obtaining the PDF and the corresponding Fokker-Planck equation can also be applied to the fractional Langevin equation (5.74) for linear force.

From Eq. (5.78) one can obtain the velocity $v(t)$ which is given by

$$v(t) = \langle v \rangle + \int_0^t dt_1 g_\alpha \left(t - t_1\right) Q(t_1) \quad (5.80)$$

with $G_\alpha(0) = 0$, where

$$g_\alpha(t) = \frac{dG_\alpha(t)}{dt}, \tag{5.81}$$

$$\langle v \rangle = v_0 \, {}^C_0 D_t^\alpha G_\alpha(t) - x_0 \omega^2 G_\alpha(t). \tag{5.82}$$

For $\alpha = 1$ (ordinary derivative) we recover the results obtained in the previous section.

From the solutions (5.78) and (5.80) and taking into account the symmetry of the correlation function, one can obtain the explicit expressions of the variances and covariance

$$\sigma_{xx}(t) = \langle x^2 \rangle - \langle x \rangle^2 = 2 \int_0^t dt_1 G_\alpha(t_1) \int_0^{t_1} dt_2 G_\alpha(t_2) \, C(t_1 - t_2), \tag{5.83}$$

$$\sigma_{xv}(t) = \frac{1}{2} \frac{d\sigma_{xx}}{dt} = G_\alpha(t) \int_0^t dt_1 G_\alpha(t_1) \, C(t - t_1) \tag{5.84}$$

and

$$\sigma_{vv}(t) = \langle v^2 \rangle - \langle v \rangle^2 = 2 \int_0^t dt_1 g_\alpha(t_1) \int_0^{t_1} dt_2 g_\alpha(t_2) \, C(t_1 - t_2) . \tag{5.85}$$

It should be noted that the solutions (5.78), (5.80), (5.83), (5.84) and (5.85) provide very general expressions due to the fact that they do not depend explicitly on the parameter $\alpha$. In fact, they maintain the same expressions of the solutions of the ordinary derivative. The order of the fractional derivative $\alpha$ appears only in the kernel $G_\alpha(t)$, $\langle x \rangle$ and $\langle v \rangle$.

In the case of $C(t)$ proportional to the dissipative memory kernel $C(t) = c\gamma(t)$ the expressions (5.83)-(5.85) can be simplified to

$$\sigma_{xx}(t) = 2 \int_0^t dt_1 G_\alpha(t_1) \int_0^{t_1} dt_2 G_\alpha(t_2) \, C(t_1 - t_2)$$

$$= c \left[ 2I_\alpha(t) - 2 \int_0^t dt_1 G_\alpha(t_1) {}^C_0 D_{t_1}^\alpha G_\alpha(t_1) - \omega^2 I_\alpha^2(t) \right], \tag{5.86}$$

$$\sigma_{xv}(t) = \frac{1}{2} \frac{d\sigma_{xx}}{dt} = c G_\alpha(t) \left[ 1 - {}^C_0 D_t^\alpha G_\alpha(t) - \omega^2 I_\alpha(t) \right] \tag{5.87}$$

and

$$\sigma_{vv}(t) = 2 \int_0^t dt_1 g_\alpha(t_1) \int_0^{t_1} dt_2 g_\alpha(t_2) C(t_1 - t_2)$$

$$= 2c \int_0^t dt_1 g_\alpha(t_1) \left\{ -_0 D_{t_1}^\alpha g_\alpha(t_1) - \omega^2 G_\alpha(t_1) \right\}$$

$$= c \left[ -2 \int_0^t dt_1 g_\alpha(t_1) {}_0 D_{t_1}^\alpha g_\alpha(t_1) - \omega^2 G_\alpha^2(t) \right], \tag{5.88}$$

where $_0 D_t^\alpha$ is the Riemann-Liouville fractional operator defined by

$$_0 D_t^\alpha f(t) = \frac{1}{\Gamma(1-\alpha)} \int_0^t dt_1 \frac{f(t_1)}{(t-t_1)^\alpha}. \tag{5.89}$$

We note that for $\alpha = 1$ we recover the results obtained in the previous section.

Next, we consider a long-time correlation function and without external force $F(x) = 0$. We take a power-law correlation function described by

$$C(t) = C_\theta t^{-\theta}, \quad 0 < \theta < 1 \tag{5.90}$$

and the frictional memory kernel as

$$\gamma(t) = \gamma_\lambda t^{-\lambda}, \quad 0 < \lambda < 1. \tag{5.91}$$

The Laplace transform of $\gamma(t)$ is given by

$$\gamma_s(s) = \gamma_\lambda \Gamma(1-\lambda) s^{\lambda-1}. \tag{5.92}$$

From Eq. (5.77) we obtain

$$G_{s\alpha}(s) = \frac{s^{-\lambda}}{\gamma_\lambda \Gamma(1-\lambda) + s^{1+\alpha-\lambda}}. \tag{5.93}$$

The Laplace inversion of Eq. (5.93) may be written in terms of the generalized Mittag-Leffler function $E_{\alpha,\beta}(z)$ (see Appendix (5.4.1)) which is given by

$$G_\alpha(t) = t^\alpha E_{1+\alpha-\lambda,1+\alpha} \left( -\gamma_\lambda \Gamma(1-\lambda) t^{1+\alpha-\lambda} \right) \tag{5.94}$$

and

$$g_\alpha(t) = t^{\alpha-1} E_{1+\alpha-\lambda,\alpha} \left( -\gamma_\lambda \Gamma(1-\lambda) t^{1+\alpha-\lambda} \right). \tag{5.95}$$

Moreover, one can obtain explicit solutions for

$$\int_0^{t_1} dt_2 G_\alpha(t_2) C(t_1 - t_2)$$

$$= C_\theta \Gamma(1-\theta) t_1^{1+\alpha-\theta} E_{1+\alpha-\lambda,2+\alpha-\theta} \left(-\gamma_\lambda \Gamma(1-\lambda) t_1^{1+\alpha-\lambda}\right) \qquad (5.96)$$

and

$$\int_0^{t_1} dt_2 g_\alpha(t_2) C(t_1 - t_2)$$

$$= C_\theta \Gamma(1-\theta) t_1^{\alpha-\theta} E_{1+\alpha-\lambda,1+\alpha-\theta} \left(-\gamma_\lambda \Gamma(1-\lambda) t_1^{1+\alpha-\lambda}\right). \qquad (5.97)$$

The explicit solutions for $\langle x \rangle$ and $\langle v \rangle$ can be obtained from the solution (5.94), and they are given by

$$\langle x \rangle = x_0 + v_0 t E_{1+\alpha-\lambda,2} \left(-\gamma_\lambda \Gamma(1-\lambda) t^{1+\alpha-\lambda}\right) \qquad (5.98)$$

and

$$\langle v \rangle = v_0 E_{1+\alpha-\lambda} \left(-\gamma_\lambda \Gamma(1-\lambda) t^{1+\alpha-\lambda}\right). \qquad (5.99)$$

We note that these first moments can give the same results obtained from the fractional Fokker-Planck equation by identifying the parameter of the fractional order of the fractional Fokker-Planck equation with $1 + \alpha - \lambda$ (for $\lambda \neq \alpha$) [27, 28]. For $\alpha = 1$ we recover the results of the generalized Langevin equation (5.1). Moreover, for $\lambda = \alpha$ we obtain the same results of the normal Brownian motion.

The asymptotic behaviors of the above quantities can be obtained by using the long-time limit of the generalized Mittag-Leffler function (see Eq. (5.112), Appendix (5.4.1))

$$E_{\alpha,\beta}(z) \sim -\frac{1}{z\Gamma(\beta-\alpha)}, \qquad (5.100)$$

and we obtain

$$\langle x \rangle \sim \frac{v_0 t^{\lambda-\alpha}}{\gamma_\lambda \Gamma(1-\lambda)\Gamma(1+\lambda-\alpha)} \qquad (5.101)$$

and

$$\langle v \rangle \sim \frac{v_0 t^{\lambda-\alpha-1}}{\gamma_\lambda \Gamma(1-\lambda)\Gamma(\lambda-\alpha)}, \qquad (5.102)$$

where we have considered $x_0 = 0$. Eq. (5.101) shows a net drift in a direction determined by the initial velocity $v_0$. For $\lambda < \alpha$, Eq. (5.101) exhibits a slow power law decay. We see that there is a competition between the dissipative term and inertial term. In the normal Brownian motion $\langle v \rangle$ exhibits an exponential decay faster than the power law decay exhibited by Eq. (5.102) for $\lambda \neq \alpha$ ($0 < \lambda < 1$ and $0 < \alpha < 1$). But, for $\lambda = \alpha$, $\langle v \rangle$ exhibits an exponential decay which has the same result of the normal Brownian motion.

Instead of determining the complete solutions of the variances by using (5.83)-(5.85), we are interested in determining the long-time behaviors of $\sigma_{xx}$ which can be obtained from Eq. (5.83). In this case, $\sigma_{xx}$ is given by

$$\sigma_{xx}(t) \sim \text{const.}, \quad 2\lambda < \theta , \tag{5.103}$$

$$\sigma_{xx}(t) \sim \ln(t), \quad 2\lambda = \theta \tag{5.104}$$

and

$$\sigma_{xx}(t) \sim t^{2\lambda - \theta}, \quad 2\lambda > \theta . \tag{5.105}$$

From Eq. (5.105) we have: the normal diffusion for $2\lambda - \theta = 1$, subdiffusion $2\lambda - \theta < 1$ and superdiffusion $2\lambda - \theta > 1$. It is interesting to note that the long-time behavior of $\sigma_{xx}(t)$ does not depend on the parameter $\alpha$ of the inertial term. This means that the inertial term has no significant contribution to the long-time limit of $\sigma_{xx}(t)$ as in the ordinary case [19]. It seems that the inertial term $dv/dt$ replaced by a fractional derivative promotes changes in the initial- and intermediate-time behaviors of the generalized Langevin equation (5.1), when $x_0 = v_0 = 0$. This change may be used to improve the data fittings of the generalized Langevin equation (5.1).

Other generalizations of the Langevin equation have also been considered, for instance, the Langevin equation with two fractional orders [29]; fractional Langevin equations with a different interpretation for the velocity [24, 25].

## 5.4 Appendices

### 5.4.1 *Generalized Mittag-Leffler function*

The generalized Mittag-Leffler function in two parameters is defined by [30]

$$E_{\mu,\nu}(z) = \sum_{n=0}^{\infty} \frac{z^n}{\Gamma(\nu + \mu n)}, \quad \mu > 0, \ \nu > 0, \tag{5.106}$$

and it reduces to the Mittag-Leffler function $E_{\mu}(z)$ by setting $\nu = 1$. The following relations follow from the definition (5.106):

$$E_{1,1}(z) = \sum_{n=0}^{\infty} \frac{z^n}{\Gamma(1+n)} = e^z, \tag{5.107}$$

$$E_{1,2}(z) = \frac{e^z - 1}{z}, \tag{5.108}$$

$$E_{1,3}(z) = \frac{e^z - 1 - z}{z^2}, \tag{5.109}$$

$$E_{2,1}(z^2) = \cosh(z), \tag{5.110}$$

$$E_{2,2}(z^2) = \frac{\sinh(z)}{z}. \tag{5.111}$$

The function $E_{\mu,\nu}(-z)$, $z \in R^+$, is completely monotonic for $0 < \mu \leq 1$ and $\nu > \mu$ [14]. In particular, for $z \to \infty$ we have [30]

$$E_{\mu,\nu}(z) \sim -\sum_{n=1}^{N} \frac{z^{-n}}{\Gamma(\nu - \mu n)}, \quad |\arg(-z)| < \left(1 - \frac{\alpha}{2}\right)\pi. \tag{5.112}$$

The $n$-th derivative of the Mittag-Leffler function is given by

$$E_{\mu,\nu}^{(n)}(z) = \frac{d^n}{dy^n} E_{\mu,\nu}(z) = \sum_{k=0}^{\infty} \frac{(k+n)! z^k}{k! \Gamma(\nu + \mu(k+n))}. \tag{5.113}$$

Eq. (5.113) is obtained as follows. Applying the $n$-th derivative to Eq. (5.106) yields

$$E_{\mu,\nu}^{(n)}(z) = \sum_{j=n}^{\infty} \frac{j(j-1)...(j-n+1) z^{j-n}}{\Gamma(\nu + \mu j)}. \tag{5.114}$$

Replacing $k = j - n$ we obtain

$$E_{\mu,\nu}^{(n)}(z) = \sum_{k=0}^{\infty} \frac{(k+n)(k+n-1)...(k+1) z^k}{\Gamma(\nu + \mu(k+n))}. \tag{5.115}$$

Eq. (5.115) can be written as

$$E_{\mu,\nu}^{(n)}(z) = \sum_{k=0}^{\infty} \frac{(k+n)(k+n-1)...(k+1) \times k!z^k}{k!\Gamma(\nu+\mu(k+n))}$$

$$= \sum_{k=0}^{\infty} \frac{(k+n)!z^k}{k!\Gamma(\nu+\mu(k+n))}. \tag{5.116}$$

We see that the Mittag-Leffler function has a power-law decay and it is difficult to obtain its asymptotic values numerically. The following procedure can be adopted to obtain all values of the Mittag-Leffler function. We first apply the Laplace transform to the Mittag-Leffler function, and it is given by

$$\mathcal{L}[E_{\mu}(-\lambda t^{\mu})] = \frac{s^{\mu-1}}{\lambda+s^{\mu}}, \quad \mathrm{Re}\ s > |\lambda|^{\frac{1}{\mu}}. \tag{5.117}$$

The numerical computation of the Mittag-Leffler function can be performed by using the numerical inversion formula given in Appendix 4.6.1. The procedure can also be applied to the generalized Mittag-Leffler function; the Laplace transform of the generalized Mittag-Leffler function is given by

$$\mathcal{L}[t^{\nu-1}E_{\mu,\nu}(-\lambda t^{\mu})] = \frac{s^{\mu-\nu}}{\lambda+s^{\mu}}, \quad \mathrm{Re}\ s > |\lambda|^{\frac{1}{\mu}}. \tag{5.118}$$

## Bibliography

[1] R. Kubo, M. Toda and N. Hashitsume, *Statistical Physics II: Nonequilibrium Statistical Mechanics* (Springer, Berlin, 1985).

[2] B. J. West and S. Picozzi, *Phys. Rev. E* **65**, 037106 (2002).

[3] S. Picozzi and B. J. West, *Phys. Rev. E* **66**, 046118 (2002).

[4] K. G. Wang, *Phys. Rev. A* **45**, 833 (1992).

[5] J. M. Porra, K. G. Wang and J. Masoliver, *Phys. Rev. E* **53**, 5872 (1996).

[6] K. G. Wang and M. Tokuyama, *Physica A* **265**, 341 (1999).

[7] A. D. Viñales and M. A. Despósito, *Phys. Rev. E* **73**, 016111 (2006).

[8] G. A. Pavliotis, *Stochastic Processes and Applications* (Springer, New York, 2014).

[9] K. S. Fa, *Phys. Rev E* **73**, 061104 (2006).

[10] K. S. Fa, *J. Math. Phys.* **50**, 083301 (2009).

[11] A. Taloni, A. Chechkin and J. Klafter, *Phys. Rev. Lett* **104**, 160602 (2010).

[12] L. Kantorovich, *Phys. Rev. B* **78**, 094304 (2008).

[13] L. Stella, C. D. Lorenz and L. Kantorovich, *Phys. Rev. B* **89**, 134303 (2014).

[14] R. Gorenflo and F. Mainardi, "Fractional calculus: Integral and differential equations of fractional order" in *Fractals and Fractional Calculus in Continuum Mechanics*, edited by A. Carpinteri and F. Mainardi (Springer, Wien, 1997), pp. 223-276.

[15] S. Kawai and T. Komatsuzaki, *J. Chem. Phys.* **134**, 114523 (2011).

[16] T. Franosch1, M. Grimm, M. Belushkin, F. M. Mor, G. Foffi, L. Forro and S. Jeney, *Nature* **478**, 85 (2011).

[17] K. R. Symon, *Mechanics*, third edition (Addison-Wesley, Philippines, 1971).

[18] H. Goldstein, C. Poole and J. Safko, *Classical Mechanics*, third edition (Pearson, USA, 2014).

[19] H. Risken, *The Fokker-Planck Equation*, second ed. (Springer-Verlag, Berlin, 1996).

[20] A. D. Viñales and M. A. Despósito, *Phys. Rev. E* **75**, 042102 (2007).

[21] R. F. Grote and J. T. Hynes, *J. Chem. Phys.* **73**, 2715 (1980).

[22] A. Neiman and W. Sung, *Phys. Lett. A* **223**, 341 (1996).

[23] J. D. Bao and Y. Z. Zhuo, *Phys. Rev. Lett.* **91**, 138104 (2003).

[24] A. Liemert, T. Sandev and H. Kantz, *Physica A* **466**, 356 (2017).

[25] V. Kobelev and E. Romanov, *Prog. Theor. Phys. Suppl.* **139**, 470 (2000).

[26] S. I. Denisov and W. Horsthemke, *Phys. Rev. E* **62**, 7729 (2000).

[27] R. Metzler and J. Klafter, *J. Phys. Chem. B* **104**, 3851 (2000).

[28] E. Barkai and R. J. Silbey, *J. Phys. Chem. B* **104**, 3866 (2000).

[29] S. C. Lim, M. Li and L. P. Teo, *Phys. Lett A* **372**, 6309 (2008).

[30] A. Erdélyi, W. Magnus, F. Oberhettinger and F. G. Tricomi, *Higher Transcendental Functions*, Bateman Project, Vol. 3 (McGraw-Hill, New York, 1955).

## Additional references for chapters 2-5

[1] P. Hänggi, P. Talkner and M. Borkovec, *Rev. Mod. Phys.* **62**, 251 (1990).

[2] P. Jung, *Phys. Rep.* **234**, 175 (1993).

[3] B. J. West and W. Deering, *Phys. Rep.* **246**, 1 (1994).

[4] P. Fröbrich and I. I. Gontchar, *Phys. Rep.* **292**, 131 (1998).

[5] L. Gammaitoni, P. Hänggi, P. Jung and F. Marchesoni, *Rev. Mod. Phys.* **70**, 223 (1998).

[6] P. S. Landa and P. V. E. McClintock, *Phys. Rep.* **323**, 1 (2000).

[7] F. So and K. L. Liu, *Physica A* **277**, 335 (2000).

[8] F. So and K. L. Liu, *Physica A* **303**, 79 (2002).

[9] P. Reimann, *Phys. Rep.* **361**, 57 (2002).

[10] P. Reimann and P. Hänggi, *Appl. Phys. A* **75**, 169 (2002).

[11] B. Lindner, J. Garcia-Ojalvo, A. Neiman and L. Schimansky-Geier, *Phys. Rep.* **392**, 321 (2004).

[12] K. S. Fa, *Braz. J. Phys.* **36**, 777 (2006).

[13] T. Srokowski, *Phys. Rev. E* **80**, 051113 (2009).

[14] R. Huang, I. Chavez, K. M. Taute, B. Lukić, S. Jeney, M. G. Raizen and E. L. Florin, *Nature Phys.* **7**, 576 (2011).

[15] M. Li and W. Zhao, *Math. Probl. Eng.*, 673648 (2012).

[16] T. Srokowski, *Phys. Rev. E* **89**, 030102 (2014).

[17] X. Bian, C. Kim and G. E. Karniadakis, *Soft Matt.* **12**, 6331 (2016).

# Chapter 6

# Continuous Time Random Walk model

## 6.1 Introduction

The continuous-time random walk (CTRW) [1] model is a useful tool for the description of diffusion in nonequilibrium systems [2–9], which is broadly applied to natural systems and life sciences such as a particle's motion in biological systems, financial stock markets and geosciences. The CTRW model is based on the length of a given jump associated with the waiting time elapsing between two successive jumps, and these quantities are connected by a probability density function (PDF) $\psi(x,t)$ of jumps. Besides, the CTRW model may be described by a set of Langevin equations [2,10,11] or an appropriate generalized master equation [6,12,13]. The PDF $\rho(x,t)$, without the presence of external force, obeys the following equation in the space-time domain [2,12,14]:

$$\rho(x,t) = \Psi(t)\delta(x) + \int_0^t \int_{-\infty}^\infty \psi\left(x - x^{'}, t - t^{'}\right) \rho(x^{'}, t^{'}) \mathrm{d}x^{'} \mathrm{d}t^{'}, \qquad (6.1)$$

where $\delta(x)$ is the Dirac delta function, $g(t)$ is the waiting time PDF defined by

$$g(t) = \int_{-\infty}^\infty \psi(x,t)\mathrm{d}x, \qquad (6.2)$$

and $\Psi(t)$ is the cumulative probability defined by

$$\Psi(t) = 1 - \int_0^t g(\tau)d\tau. \qquad (6.3)$$

In Fourier-Laplace space the corresponding PDF is described by

$$\rho_{ks}(k,s) = \frac{1 - g_s(s)}{s} \frac{\rho_{k0}(k)}{1 - \psi_{ks}(k,s)}, \qquad (6.4)$$

where $\rho_{k0}(k)$ is the Fourier transform of the initial condition $\rho_0(x)$ and $\psi_{ks}(k,s)$ is the Fourier-Laplace transform of the jump PDF $\psi(x,t)$.

As examples of the waiting time PDF we consider the following systems:

1) The mechanochemical perturbation can affect the cells such as the lipid bilayer properties. The effect of membrane tension on lipid lateral diffusion in the dioleoylphosphocholine (DOPC) bilayer has been analyzed by molecular dynamics simulations [15]. Fig. 6.1 shows the waiting time distributions of the lipid molecules at membrane tension values of 0 dyn/cm and 15 dyn/cm obtained by using random walk analysis; the data are nicely fitted to a sum of two exponentials.

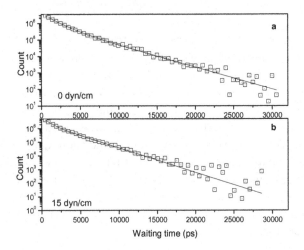

Fig. 6.1    Best fitted sum of two exponentials $c_0 \exp(a_0 t) + c_1 \exp(a_1 t)$ of the simulation result calculated in Ref. [15] at membrane tension values of 0 dyn/cm and 15 dyn/cm.

2) Another system related to the waiting time distribution is the tick-by-tick dynamics of markets. Analyses of different waiting time scales in different financial markets have shown that the data display non-exponential forms [16–21]. An example of the data is the waiting time distribution for BUND futures prices traded at the London International Financial Futures and options Exchange (LIFFE) in 1997 [16, 17]. BUND (it is a German word for bond) futures is a futures contract that obliges the holder to buy or sell a bond at maturity. Usually, transactions begin some months before the delivery date, and they begin with a few trades a day. Besides, in the financial market applications, the price of an asset and the times between two trades are considered as random variables. The variable $x$ represents

the log-price given by $x = \log S$, where $S$ is the price of an asset at time $t$. In contrast, times between two trades are represented by the waiting times. Figs. 6.2 and 6.3 show the survival probability for BUND futures traded at LIFFE with delivery date: June and September of 1997. The data are well fitted to a sum of stretched exponentials given by

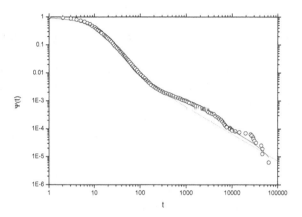

Fig. 6.2  Survival probability for BUND futures traded at LIFFE with delivery date: June of 1997. Solid line is obtained from Eq. (6.5) and the dotted line is obtained from the Mittag-Leffler function $E_{\alpha,1}(-(\lambda t)^{\alpha})$.

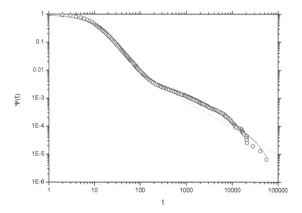

Fig. 6.3  Survival probability for BUND futures traded at LIFFE with delivery date: September of 1997. Solid line is obtained from Eq. (6.5) and the dotted line is obtained from the Mittag-Leffler function $E_{\alpha,1}(-(\lambda t)^{\alpha})$.

$$\Psi(t) = \frac{\sum_{i=0}^{n} c_i e^{-(\lambda_i t)^{\alpha_i}}}{\sum_{j=0}^{n} c_j} . \tag{6.5}$$

The survival probability corresponds to the following waiting time PDF:

$$g(t) = \frac{\sum_{i=0}^{n} c_i \alpha_i \lambda_i^{\alpha_i} t^{\alpha_i - 1} e^{-(\lambda_i t)^{\alpha_i}}}{\sum_{j=0}^{n} c_j} . \tag{6.6}$$

## 6.2   Uncoupled Continuous Time Random Walk model

The CTRW model for continuous displacement per step is based on the length of a given instantaneous jump from one site to another associated with the waiting time elapsing between two successive jumps. The waiting time on a site is described by a waiting time PDF $g(t)$ and it is independent of the waiting time of any subsequent step. Now we consider a one-dimensional process starting at $x = 0$ and $t = 0$, and $t_n$ is the waiting time for the $n$-th step of the random walker. The probability density of having a particle at $x$ after $n$ steps is denoted by $\lambda_n(x)$ with $\lambda_0(x) = \delta(x)$ and $\lambda_1(x) = \lambda(x)$, and $g_n(t)$ is the probability density of the occurrence of $n$-th step at time $t = t_1 + t_2 + ... + t_{n-1} + t_n$ with $g_0(t) = \delta(t)$ and $g_1(t) = g(t)$. For independent steps and waiting times the probability densities $\lambda_n(x)$ and $g_n(t)$ have the following recursion relations:

$$\lambda_n(x) = \int_{-\infty}^{\infty} \lambda_{n-1}(y)\lambda(x - y)dy \tag{6.7}$$

and

$$g_n(t) = \int_0^t g_{n-1}(\tau)g(t - \tau)d\tau . \tag{6.8}$$

The Fourier transform of the iteration of the relation (6.7) leads to the following expression:

$$\lambda_{kn}(k) = \lambda_k^n(k) , \tag{6.9}$$

where the subscript $k$ denotes the function $\lambda_n(x)$ in Fourier space; whereas the Laplace transform of the iteration of the relation (6.8) leads to the following expression:

$$g_{sn}(s) = g_s^n(s) , \tag{6.10}$$

where the subscript $s$ denotes the function $g_n(t)$ in Laplace space. The probability of taking exactly $n$ steps up to the time $t$ is given by

$$\xi_n(t) = \int_0^t g_n(\tau)\Psi(t - \tau)d\tau , \tag{6.11}$$

where $\Psi(t)$ is the survival probability on a site given by

$$\Psi(t) = 1 - \int_0^t g\left(t'\right) dt' = \int_t^\infty g\left(t'\right) dt' \qquad (6.12)$$

or

$$g(t) = -\frac{d\Psi(t)}{dt}. \qquad (6.13)$$

The survival probability assigns to the probability of no jump event occurs during the interval $(0, t)$. In Laplace space the probability $\xi_n(t)$ turns out to be

$$\xi_{sn}(s) = g_s^n(s)\Psi_s(s) = g_s^n(s)\frac{1 - g_s(s)}{s} . \qquad (6.14)$$

The PDF of the position of a random walker at time $t$ is given by

$$\rho(x, t) = \sum_{n=0}^\infty \lambda_n(x)\xi_n(t) . \qquad (6.15)$$

Applying the Fourier-Laplace transform to Eq. (6.15) yields

$$\rho_{ks}(k, s) = \sum_{n=0}^\infty \lambda_{kn}(k)\xi_{sn}(s) . \qquad (6.16)$$

Substituting the relations (6.9) and (6.14) into (6.16) leads to the following expression for $\rho_{ks}(k, s)$:

$$\rho_{ks}(k, s) = \frac{1 - g_s(s)}{s}\sum_{n=0}^\infty \lambda_k^n(k)g_s^n(s) = \frac{1 - g_s(s)}{s}\frac{1}{1 - \lambda_k(k)g_s(s)} . \qquad (6.17)$$

Eq. (6.17) is the main result of the uncoupled CTRW model in one dimension. For more than one dimension, the PDF to find a random walker at position $\mathbf{r}$ and time $t$ is described by

$$\rho_{ks}(\mathbf{k}, s) = \frac{1 - g_s(s)}{s}\frac{1}{1 - \lambda_k(\mathbf{k})g_s(s)} . \qquad (6.18)$$

## 6.3 Integro-differential equations

In this section the integro-differential Fokker-Planck equation and integro-differential Klein-Kramers equation are derived asymptotically in space from the decoupled continuous-time random walk model and a generalized Chapman-Kolmogorov equation, with generic waiting time probability density function $g(t)$ and external force. The integro-differential equations can be used to investigate the entire diffusion process i.e., covering initial-, intermediate-, and long-time ranges of the process, i.e, this approach can distinguish the evolution detail for a system having the same behavior in the long-time limit but with different initial- and intermediate-time behaviors.

### 6.3.1    *Integro-differential diffusion equation and Integro-differential Fokker-Planck equation*

In particular, the CTRW model (6.17) can be classified by the characteristic waiting time $\Lambda$ and the jump length variance $\Sigma^2$ defined by

$$\Lambda = \int_0^\infty t g(t) dt, \tag{6.19}$$

and

$$\Sigma^2 = \int_{-\infty}^\infty x^2 \lambda(x) dx. \tag{6.20}$$

For finite $\Lambda$ and $\Sigma^2$, the long-time limit corresponds to the Brownian motion [2]. Under the case of finite jump length variance [2] we can take the following asymptotic approximation:

$$\lambda_k(k) \sim 1 - Ck^2 , \tag{6.21}$$

where $\sqrt{C}$ has a dimension of length; the PDF for the CTRW model (6.17) is now given by

$$\rho_{ks}(k, s) = \frac{(1 - g_s(s)) \rho_{k0}(k)}{s \left[1 - (1 - Ck^2) g_s(s)\right]} \tag{6.22}$$

in Laplace-Fourier space, where $\rho_{k0}(k)$ is the Fourier transform of the initial condition $\rho_0(x)$. Although Eq. (6.22) is valid for a finite jump length variance, anomalous diffusion processes can be obtained from Eq. (6.22) with appropriate choices of $g(t)$. In particular, for a long-tailed power-law waiting time PDF $g(t) \sim (\tau/t)^{1+\alpha}$ with $0 < \alpha < 1$, Eq. (6.22) can be linked to the following fractional diffusion equation:

$$\frac{\partial \rho(x, t)}{\partial t} = {_0}D_t^{1-\alpha} K_\alpha \frac{\partial^2}{\partial x^2} \rho(x, t), \tag{6.23}$$

where $K_\alpha$ is the generalized diffusion constant, ${_0}D_t^{1-\alpha}$ is the Riemann-Liouville fractional operator defined by

$${_0}D_t^{1-\alpha} f(t) = \frac{1}{\Gamma(\alpha)} \frac{d}{dt} \int_0^t dt_1 \frac{f(t_1)}{(t - t_1)^{1-\alpha}} , \quad 0 < \alpha < 1, \tag{6.24}$$

and $\Gamma(z)$ is the Gamma function. Eq. (6.23) describes subdiffusion process for all the time intervals [2]. The derivation of Eq. (6.23) is obtained as follows. The asymptotic long-tailed power-law waiting time PDF $g(t) \sim (\tau/t)^{1+\alpha}$ has the following corresponding expression in Laplace space:

$$g_s(s) \sim 1 - (s\tau)^\alpha. \tag{6.25}$$

Substituting Eq. (6.25) into Eq. (6.22) we obtain

$$[s\rho_{ks}(k,s) - \rho_{k0}(k)] = -\frac{Csk^2\rho_{ks}(k,s)}{(s\tau)^\alpha}. \tag{6.26}$$

We first apply the inverse Fourier transform to Eq. (6.26), and we find

$$[s\rho_s(x,s) - \rho(x,0)] = C\frac{s}{(s\tau)^\alpha}\frac{\partial^2\rho_s(x,s)}{\partial x^2}. \tag{6.27}$$

Eq. (6.23) is obtained by applying the inverse Laplace transform and the convolution theorem to Eq.(6.27), with $K_\alpha = C/\tau^\alpha$.

Usually, analysis of diffusion processes is often restricted to a long-time limit. However, the information in the regime of the initial and intermediate times is very important to distinguish the differences for the systems with the same behavior in the long-time limit. For any waiting time PDF $g(t)$, Eq. (6.22) can be connected to the following integro-differential diffusion equation:

$$\frac{\partial\rho(x,t)}{\partial t} - \int_0^t dt_1 g(t - t_1)\frac{\partial\rho(x,t_1)}{\partial t_1}$$

$$= C\frac{\partial}{\partial t}\int_0^t dt_1 g(t - t_1)\frac{\partial^2\rho(x,t_1)}{\partial x^2}. \tag{6.28}$$

The derivation of Eq. (6.28) is obtained as follows. We first apply the inverse Fourier transform to Eq. (6.22), and we find

$$s\rho_s(x,s) - \rho(x,0) - g_s(s)[s\rho_s(x,s) - \rho(x,0)] = Csg_s(s)\frac{\partial^2\rho_s(x,s)}{\partial x^2}. \tag{6.29}$$

Eq. (6.28) is obtained by applying the inverse Laplace transform and the convolution theorem to Eq.(6.29). Eq. (6.28) can also be written in a compact form. Note that Eq. (6.29) can be written as

$$s\rho_s(x,s) - \rho(x,0) = C\frac{sg_s(s)}{1 - g_s(s)}\frac{\partial^2\rho_s(x,s)}{\partial x^2}. \tag{6.30}$$

Applying the inverse Laplace transform and the convolution theorem to Eq.(6.30) yields

$$\frac{\partial\rho(x,t)}{\partial t} = C\frac{\partial}{\partial t}\int_0^t dt_1 g^*(t - t_1)\frac{\partial^2\rho(x,t_1)}{\partial x^2}, \tag{6.31}$$

where $g^*(t)$ is defined through its Laplace transform as follows:

$$g_s^*(s) = \frac{g_s(s)}{1 - g_s(s)}. \tag{6.32}$$

In the presence of an external force, the integro-differential diffusion equation can be extended [22, 23] to

$$\frac{\partial \rho(x,t)}{\partial t} - \int_0^t dt_1 g\left(t - t_1\right) \frac{\partial \rho(x, t_1)}{\partial t_1}$$

$$= C \frac{\partial}{\partial t} \int_0^t dt_1 g\left(t - t_1\right) L_{FP} \rho(x, t_1), \tag{6.33}$$

where

$$L_{FP} = -\frac{\partial}{\partial x} \frac{F(x)}{k_B T} + \frac{\partial^2}{\partial x^2}, \tag{6.34}$$

$k_B$ is the Boltzmann constant and $T$ is the absolute temperature. Eq. (6.33) is called the integro-differential Fokker-Planck equation; it can also be viewed as an integro-differential diffusion equation with the presence of an external force. The derivation of Eq. (6.33) appears in Appendix 6.4.1. The compact form of Eq. (6.33) is written as

$$\frac{\partial \rho(x,t)}{\partial t} = C \frac{\partial}{\partial t} \int_0^t dt_1 g^* \left(t - t_1\right) L_{FP} \rho(x, t_1). \tag{6.35}$$

It is important to note that Eq. (6.35) recovers the ordinary Fokker-Planck equation for the exponential waiting time PDF. However, Eq. (6.35) reduces to the fractional diffusion equation (6.23) only if $F(x) = 0$, and the waiting time PDF is given by

$$g(t) = -\frac{d}{dt} E_{\alpha,1}\left(-\lambda_\alpha t^\alpha\right) = \lambda_\alpha t^{\alpha-1} E_{\alpha,\alpha}(-\lambda_\alpha t^\alpha), 0 < \alpha \leq 1, \tag{6.36}$$

where $E_{\mu,\nu}(y)$ is the generalized Mittag-Leffler function [24, 25] (see Appendix 5.4.1).

In contrast to the asymptotic approximation of the finite jump length variance given by (6.21) the Lévy flights are characterized by the diverging jump length variance corresponding to the following asymptotic power-law form [26, 27]:

$$\lambda(x) \sim \frac{C_\mu}{|x|^{1+\mu}}. \tag{6.37}$$

The Lévy distribution for the jump length in Fourier space [28] is given by

$$\lambda_k(k) = \exp(-C_\mu |k|^\mu). \tag{6.38}$$

For small $k$ it has the following approximation:

$$\lambda_k(k) \sim 1 - C_\mu |k|^\mu. \tag{6.39}$$

A fractional diffusion equation related to the asymptotic jump length (6.39) can be obtained by substituting Eq. (6.39) into (6.17), and it is given by

$$\frac{\partial \rho(x,t)}{\partial t} = C_\mu \frac{\partial}{\partial t} \int_0^t \mathrm{d}t_1 g^* (t - t_1) \frac{\partial^\mu \rho(x,t_1)}{\partial |x|^\mu} , \qquad (6.40)$$

where $C_\mu^{1/\mu}$ has a dimension of length and $\frac{\partial^\mu}{\partial |x|^\mu}$ is the Riesz space fractional derivative (see Appendix 6.4.2 for the definition of the Riesz space fractional derivative and its Fourier transform). Eq. (6.40) reduces to the fractional diffusion equation both in space and time [2, 29, 30] for $g(t)$ given by Eq. (6.36),

$$\frac{\partial \rho(x,t)}{\partial t} =_0 D_t^{1-\alpha} K_{\alpha\mu} \frac{\partial^\mu \rho(x,t)}{\partial |x|^\mu} , \qquad (6.41)$$

where

$$K_{\alpha\mu} = \frac{C_\mu}{\tau^\alpha}. \qquad (6.42)$$

### 6.3.2 Generalized Chapman-Kolmogorov equation and integro-differential Klein-Kramers equation

Eq. (6.33) only provides the PDF in position space, i.e., $\rho(x,t)$. In classic approach, the PDF $\rho(x,v,t)$ satisfies the ordinary Klein-Kramers equation, which was derived by Kramers to describe reaction kinetics [31]; the equation has also been used to describe, for instance, transport in superionic conductors, Josephson tunneling junctions and relaxation of dipoles [32]. For one-dimensional space the ordinary Klein-Kramers equation is given by

$$\frac{\partial \rho(x,v,t)}{\partial t} = \left[ -v\frac{\partial}{\partial x} - \frac{F(x)}{m}\frac{\partial}{\partial v} + \gamma L_R \right] \rho(x,v,t) \qquad (6.43)$$

and

$$L_R = \frac{\partial}{\partial v}v + \frac{k_B T}{m}\frac{\partial^2}{\partial v^2} , \qquad (6.44)$$

where $m$ is the mass of Brownian particle.

In order to find $\rho(x,v,t)$ for the CTRW model, we express it in terms of the following generalized Chapman-Kolmogorov equation as follows [33,34]:

$$\rho(x,v,t) = \Psi(t)\rho(x,v,0) + \int_0^t \mathrm{d}t_1 \int_{-\infty}^\infty \mathrm{d}\Delta x \int_{-\infty}^\infty \mathrm{d}\Delta v$$

$$\times \rho(x - \Delta x, v - \Delta v, t_1)\Phi(x - \Delta x, v - \Delta v; \Delta v) g(t - t_1), \qquad (6.45)$$

where the survival probability $\Psi(t)$ expresses the persistence of the initial condition $\rho(x, v, 0)$,

$$\Phi(x - \Delta x, v - \Delta v; \Delta v) = \phi(x - \Delta x, v - \Delta v; \Delta v)\, \delta(\Delta x - vt) \qquad (6.46)$$

is the transition distribution, $\Delta x$ and $\Delta v$ are position and velocity increments. By specifying the waiting time PDF as $g(t) = \delta(t - \Delta t)$, where $\Delta t$ is the jump time, the ordinary Klein-Kramers equation can be obtained from Eq. (6.45). For any other kind of waiting time PDF the trapping events should be included in the system. During a trapping event the particle is temporarily immobilized which is described by the waiting time PDF. Therefore, for short times less than a cut-off time $\tau^*$ ( $t < \tau^*$) the test particle moves with velocity $v$ in a given direction, and the distance covered by the particle is $\Delta x = vt$. Whereas, for $t > \tau^*$ the particle is temporarily immobilized at the current position associated with the waiting time PDF, and then the covered distance during this trapping event is kept constant that is given by $\Delta x = v\tau^*$. This mechanism of trapping event results in that Eq. (6.45) can be split as

$$\rho(x, v, t)$$

$$= \int_0^{\tau^*} dt_1 \int_{-\infty}^{\infty} d\Delta v \rho(x - vt_1, v - \Delta v, t_1)\phi(x - vt_1, v - \Delta v; \Delta v)\, g(t - t_1)$$

$$+ \int_{\tau^*}^{t} dt_1 \int_{-\infty}^{\infty} d\Delta v \rho(x - v\tau^*, v - \Delta v, t_1)\phi(x - v\tau^*, v - \Delta v; \Delta v)\, g(t - t_1)$$

$$+ \Psi(t)\rho(x, v, 0). \qquad (6.47)$$

Expanding Eq. (6.47) in terms of Taylor series with respect to $x$ and $v$, and using the moments of the mean velocity increments given in Ref. [35]

$$\langle \Delta v \rangle = -\left(\eta v - \frac{F(x)}{m}\right)\Delta t \qquad (6.48)$$

and

$$\langle (\Delta v)^2 \rangle = \frac{2\eta k_B T}{m}\Delta t \qquad (6.49)$$

yields

$$\rho(x, v, t) = \int_0^t dt_1 g(t - t_1)\, \rho(x, v, t_1) + \int_0^{\tau^*} dt_1 t_1 g(t - t_1)\, L_{KK}\rho(x, v, t_1)$$

$$+ \tau^* \int_{\tau^*}^{t} dt_1 g(t - t_1)\, L_{KK}\rho(x, v, t_1) + \Psi(t)\rho(x, v, 0), \qquad (6.50)$$

where

$$L_{KK} = -v\frac{\partial}{\partial x} + \frac{\partial}{\partial v}\left(\eta v - \frac{F(x)}{m}\right) + \frac{\eta k_B T}{m}\frac{\partial^2}{\partial v^2}. \qquad (6.51)$$

Eq. (6.50) is difficult to be solved exactly due to splitting time. However, for the special case of a long-tailed power-law waiting time, Eq. (6.50) can approximately be replaced by the fractional Klein-Kramers equation [33,34] given by

$$\frac{\partial\rho(x,v,t)}{\partial t} = {}_0D_t^{1-\alpha}\left[-v^*\frac{\partial}{\partial x} - \frac{F^*(x)}{m}\frac{\partial}{\partial v} + \eta^*L_R\right]\rho(x,v,t) , \qquad (6.52)$$

where the starred quantities are defined in terms of the inter-trapping time scale $\tau^*$ and the internal waiting time scale $\tau$ as $v^* = v\vartheta$, $\eta^* = \eta\vartheta$, $F^*(x) = F(x)\vartheta$ and $\vartheta = \tau^*/\tau^\alpha$. Note that Eqs. (6.23) and (6.52) describe subdiffusive regime for force-free case.

For generic waiting time PDF Eq. (6.50) can be rewritten as

$$\rho(x,v,t) = \Psi(t)\rho(x,v,0)$$

$$+ \int_0^t dt_1 g\left(t - t_1\right)\rho(x,v,t_1) + \int_0^t dt_1\theta\left(t_1\right)g\left(t - t_1\right)L_{KK}\rho(x,v,t_1), \qquad (6.53)$$

where

$$\theta\left(t\right) = \tau^* + \left(t - \tau^*\right)\Theta\left(\tau^* - t\right) \qquad (6.54)$$

and $\Theta\left(t\right)$ is the Heaviside function. We now consider the following first mean-value theorem [36]: $\int_{q_1}^{q_2} dx f(x)h(x) = f(\xi)\int_{q_1}^{q_2} dx h(x)$ , where $f(x)$ is monotonic and non-negative throughout the interval $(q_1,q_2)$, with $q_1 \leq \xi \leq q_2$, and $h(x)$ is an integrable function. Using the first mean-value theorem, we obtain from Eq. (6.53) the following result:

$$\rho(x,v,t) = \Psi(t)\rho(x,v,0)$$

$$+ \int_0^t dt_1 g\left(t - t_1\right)\rho(x,v,t_1) + \tau_i\int_0^t dt_1 g\left(t - t_1\right)L_{KK}\rho(x,v,t_1), \qquad (6.55)$$

where $0 < \tau_i \leq \tau^*$. It should be noted that Eq. (6.55) is an exact equation of Eq. (6.53), i.e., not approximate, for PDF at a given time $t$ once $\tau_i$ is known. Since $\tau_i$ is different for different time $t$, Eq. (6.55) can be thought of an approximate equation of Eq. (6.53) with varying time $t$. The error can be reduced by choosing small cut-off time $\tau^*$. The case of small cut-off time corresponds to a particle (random walker) that moves with small spatial jump before it is temporarily immobilized; however, the immobilization time may be short or long, which is described by the waiting time PDF. This scenario has its counterpart in the CTRW approach with a finite jump length variance and any waiting time PDF. Differentiate Eq. (6.55) with respect to $t$ yields

$$\frac{\partial \rho(x,v,t)}{\partial t} = -g(t)\rho(x,v,0) + \frac{\partial}{\partial t}\int_0^t dt_1 g(t-t_1)\rho(x,v,t_1)$$

$$+\tau_i \frac{\partial}{\partial t}\int_0^t dt_1 g(t-t_1) L_{KK}\rho(x,v,t_1) = -g(t)\rho(x,v,0)$$

$$+g(0)\rho(x,v,t) + \int_0^t dt_1 \frac{\partial g(t-t_1)}{\partial t}\rho(x,v,t_1)$$

$$+\tau_i \frac{\partial}{\partial t}\int_0^t dt_1 g(t-t_1) L_{KK}\rho(x,v,t_1) = -g(t)\rho(x,v,0)$$

$$+y(0)\rho(x,v,t) - \int_0^t dt_1 \frac{\partial g(t-t_1)}{\partial t_1}\rho(x,v,t_1)$$

$$+\tau_i \frac{\partial}{\partial t}\int_0^t dt_1 g(t-t_1) L_{KK}\rho(x,v,t_1) = -g(t)\rho(x,v,0)$$

$$+g(0)\rho(x,v,t) - g(t-t_1)\rho(x,v,t_1)|_0^t + \int_0^t dt_1 g(t-t_1)\frac{\partial \rho(x,v,t_1)}{\partial t_1}$$

$$+\tau_i \frac{\partial}{\partial t}\int_0^t dt_1 g(t-t_1) L_{KK}\rho(x,v,t_1). \tag{6.56}$$

The last expression can be simplified, and the result is given by

$$\frac{\partial \rho(x,v,t)}{\partial t} - \int_0^t dt_1 g(t-t_1)\frac{\partial}{\partial t_1}\rho(x,v,t_1)$$

$$= \tau_i \frac{\partial}{\partial t}\int_0^t dt_1 g(t-t_1) L_{KK}\rho(x,v,t_1). \tag{6.57}$$

This is the integro-differential Klein-Kramers equation with generic waiting time PDF and external force. The compact form of Eq. (6.57) is written as

$$\frac{\partial \rho(x,v,t)}{\partial t} = \tau_i \frac{\partial}{\partial t}\int_0^t dt_1 g^*(t-t_1) L_{KK}\rho(x,v,t_1). \tag{6.58}$$

Eq. (6.58) recovers the ordinary Klein-Kramers equation for the exponential waiting time PDF, whereas for the waiting time PDF given by Eq. (6.36) it reduces to the fractional Klein-Kramers equation (6.52).

The result (6.57) can also be connected with the CTRW approach in the following way. Using the prescription of the CTRW [1, 37] we may rewrite the PDF as

$$\rho(x,v,t) = \sum_{n=0}^{\infty} \int_0^t dt_1 g_n(t_1)\Psi(t-t_1)(1+\tau_i L_{KK})^n \rho(x,v,0) , \qquad (6.59)$$

where $g_n(t)$ is given by Eq. (6.8). Applying the Laplace transform to Eq. (6.59) yields

$$\rho_s(x,v,s) = \sum_{n=0}^{\infty} [g_s(s)(1+\tau_i L_{KK})]^n \Psi_s(s)\rho(x,v,0)$$

$$= \frac{\Psi_s(s)\rho(x,v,0)}{1-g_s(s)(1+\tau_i L_{KK})} = \frac{[1-g_s(s)]\rho(x,v,0)}{s[1-g_s(s)(1+\tau_i L_{KK})]}, \qquad (6.60)$$

with $|g_s(s)(1+\tau_i L_{KK})| < 1$. Besides, the Laplace transform of Eq. (6.58) yields

$$\rho_s(x,v,s) = \frac{\rho(x,v,0)}{s[1-\tau_i g_s^*(s)L_{KK}]} = \frac{[1-g_s(s)]\rho(x,v,0)}{s[1-g_s(s)(1+\tau_i L_{KK})]} . \qquad (6.61)$$

We see that Eq. (6.60) is equal to Eq. (6.61). Thus, we have shown the equivalence between the integro-differential equation (6.57) and Eq. (6.59). Moreover, this equivalence is also valid for Eq. (6.33) by considering the following equation:

$$\rho(x,t) = \sum_{n=0}^{\infty} \int_0^t dt_1 g_n(t_1)\Psi(t-t_1)(1+CL_{FP})^n \rho(x,0) . \qquad (6.62)$$

By integrating Eq. (6.57) with respect to $x$ one can obtain that in the case of force-free the PDF $\rho(v,t)$ fulfills the following generalized Rayleigh equation :

$$\frac{\partial}{\partial t}\rho(v,t) - \int_0^t dt_1 g(t-t_1)\frac{\partial}{\partial t_1}\rho(v,t_1)$$

$$= \tau_i \eta \frac{\partial}{\partial t} \int_0^t dt_1 g(t-t_1)\left[\frac{\partial}{\partial v}v + \frac{k_B T}{m}\frac{\partial^2}{\partial v^2}\right]\rho(v,t_1) \qquad (6.63)$$

or

$$\frac{\partial \rho(v,t)}{\partial t} = \tau_i \eta \frac{\partial}{\partial t} \int_0^t dt_1 g^*(t-t_1)\left[\frac{\partial}{\partial v}v + \frac{k_B T}{m}\frac{\partial^2}{\partial v^2}\right]\rho(v,t_1) . \qquad (6.64)$$

For the initial condition $\rho(v,0) = \delta(v - v_0)$, we can show that $\rho(v,t)$, described by Eq. (6.63), is normalized.

Moreover, in the following one can prove that the PDF $\rho(x,t)$ satisfies Eq. (6.33) starting from Eq. (6.57). Taking the Laplace transform of Eq. (6.57) with respect to $t$, we obtain

$$[1 - g_s(s)]\,[s\rho_s(x,v,s) - \rho(x,v,0)] = \tau_i s g_s(s) L_{KK} \rho_s(x,v,s) . \qquad (6.65)$$

The integration of Eq. (6.65) over velocity, and of $v$ times Eq. (6.65) over velocity leads to two independent equations, and combination of these two equations with the consideration of the high friction limit yields Eq. (6.33). That both the generalized Rayleigh equation and the integro-differential Fokker-Planck equation can be derived from Eq. (6.57).

## 6.4   Appendices

### 6.4.1   *Integro-differential diffusion equation with external force*

We consider a random walk on one dimensional lattice with a lattice spacing $\bar{a}$. Once the random walk has arrived at site $n$ it is trapped there for some random time (waiting time). These waiting times are given according to the waiting time PDF $g(t)$. Therefore, the probability, $Q_i(t)$, that the random walker has jumped $i$ times in the interval $(0,t)$, is given in Laplace space by [23].

$$Q_{si}(s) = \frac{1 - g_s(s)}{s} g_s^i(s) , \qquad (6.66)$$

where $[1 - g_s(s)]/s$ corresponds to the survival probability $\Psi_s(s)$. Now let us consider $\rho_n(t)$ the probability of finding the random walker at site $n$ at time $t$, and $p_i(n)$ be the probability to be on site $n$ after step $i$. Then, we have

$$\rho_n(t) = \sum_{i=0}^{\infty} p_i(n) Q_i(t). \qquad (6.67)$$

Using Eqs. (6.66) and (6.67) we can obtain $\rho_n(t)$ in Laplace space as follows:

$$\rho_{sn}(s) = \frac{1 - g_s(s)}{s} \sum_{i=0}^{\infty} p_i(n) g_s^i(s). \qquad (6.68)$$

The evolution of $p_i(n)$ in discrete time and space is replaced by its continuum limit by $p_i(n) \to p_i(x)$, and is determined by

$$p_{i+1}(x) = R(x - \bar{a}) p_i(x - \bar{a}) + L(x + \bar{a}) p_i(x + \bar{a}), \qquad (6.69)$$

where $R(x)$ and $L(x)$ are the directional jumping probabilities with considering independent of the waiting times. By considering that the system is close to thermal equilibrium with temperature $T$, $R(x) \simeq L(x) \simeq 1/2$ and the detailed balance $R(x) - L(x) \simeq \bar{a}F(x)/(2k_BT)$ (where $F(x)$ is the external force), we now expand Eq. (6.69) in a Taylor series and omit the higher terms than $\bar{a}^2$, and we obtain

$$p_{i+1}(x) = p_i(x) + \frac{\bar{a}^2}{2}\left[\frac{\partial^2}{\partial x^2}p_i(x) - \frac{\partial}{\partial x}\frac{F(x)}{k_BT}p_i(x)\right] . \tag{6.70}$$

Considering the continuum limit $\rho_{sn}(s) \to \rho_s(x, s)$ in Eq. (6.68) and substituting Eq. (6.70) into Eq. (6.68), we obtain the following result:

$$\rho_s(x, s) = \frac{1 - g_s(s)}{s}\rho(x, 0)$$

$$+ g_s(s)\left\{1 + \frac{\bar{a}^2}{2}\left[\frac{\partial^2}{\partial x^2} - \frac{\partial}{\partial x}\frac{F(x)}{k_BT}\right]\right\}\rho_s(x, s). \tag{6.71}$$

Eq. (6.33) is obtained by applying the inverse Laplace transform to Eq. (6.71) and using the method given in Ref. [38], i.e.,

$$\frac{\partial\rho(x, t)}{\partial t} - \int_0^t g(t - t_1)\frac{\partial\rho(x, t_1)}{\partial t_1}dt_1$$

$$= CL_{FP}\frac{\partial}{\partial t}\int_0^t g(t - t_1)\rho(x, t_1)dt_1, \tag{6.72}$$

where $C = \bar{a}^2/2$.

### 6.4.2 *Riesz space fractional derivative*

The Riesz space fractional derivative is defined by [39, 40]

$$\frac{d^\alpha f(x)}{d|x|^\alpha} = -\frac{1}{2\cos(\pi\alpha/2)}\left(_{-\infty}D_x^\alpha f(x) +_x D_\infty^\alpha f(x)\right) , 0 < \alpha < 2 , \tag{6.73}$$

where

$$_{-\infty}D_x^\alpha f(x) = \frac{1}{\Gamma(n - \alpha)}\frac{d^n}{dx^n}\int_{-\infty}^x \frac{f(y)}{(x - y)^{\alpha+1-n}}dy, n - 1 < \alpha < n , \tag{6.74}$$

and

$$_x D_\infty^\alpha f(x) = \frac{(-1)^n}{\Gamma(n - \alpha)}\frac{d^n}{dx^n}\int_x^\infty \frac{f(y)}{(y - x)^{\alpha+1-n}}dy, n - 1 < \alpha < n . \tag{6.75}$$

### 6.4.2.1 Fourier transform of the Riemann-Liouville fractional integral

The Riemann-Liouville fractional integral in the interval $(-\infty, x)$ and its Fourier transform are given by

$$
_{-\infty}I_x^\alpha f(x) = {}_{-\infty}D_x^{-\alpha} f(x) = \frac{1}{\Gamma(\alpha)} \int_{-\infty}^x dy \frac{f(y)}{(x-y)^{1-\alpha}}, \ \alpha > 0, \quad (6.76)
$$

and

$$
\mathcal{F}[_{-\infty}I_x^\alpha f(x)] = \frac{1}{\Gamma(\alpha)} \int_{-\infty}^\infty dx e^{-ikx} \int_{-\infty}^x dy \frac{f(y)}{(x-y)^{1-\alpha}}, \ 0 < \alpha < 1.
$$
$$(6.77)$$

Note that Eq. (6.77) may not exist for $\alpha \geq 1$. To solve Eq. (6.77) we use the Fubini's theorem [41] $\left( \int_a^b dx \int_a^x dy f(x,y) = \int_a^b dy \int_y^b dx f(x,y) \right)$, and it is rewritten as follows:

$$
\mathcal{F}[_{-\infty}I_x^\alpha f(x)] = \frac{1}{\Gamma(\alpha)} \int_{-\infty}^\infty dy f(y) \int_y^\infty dx \frac{e^{-ikx}}{(x-y)^{1-\alpha}}. \quad (6.78)
$$

Substituting $\tau = x - y$ into Eq. (6.78) we arrive at

$$
\mathcal{F}[_{-\infty}I_x^\alpha f(x)] = \frac{f_k(k)}{\Gamma(\alpha)} \int_0^\infty d\tau \frac{e^{-ik\tau}}{\tau^{1-\alpha}} = \frac{f_k(k)}{(ik)^\alpha}. \quad (6.79)
$$

For the interval $(x, \infty)$ the Riemann-Liouville fractional integral is given by

$$
_xI_\infty^\alpha f(x) = \frac{1}{\Gamma(\alpha)} \int_x^\infty dy \frac{f(y)}{(y-x)^{1-\alpha}}. \quad (6.80)
$$

The Fourier transform of Eq. (6.80) is calculated similarly as the one of the interval $(-\infty, x)$, i.e,

$$
\mathcal{F}[_xI_\infty^\alpha f(x)] = \frac{1}{\Gamma(\alpha)} \int_{-\infty}^\infty dx e^{-ikx} \int_x^\infty dy \frac{f(y)}{(y-x)^{1-\alpha}}, 0 < \alpha < 1, \quad (6.81)
$$

and it yields

$$
\mathcal{F}[_xI_\infty^\alpha f(x)] = \frac{f_k(k)}{\Gamma(\alpha)} \int_0^\infty d\tau \frac{e^{ik\tau}}{\tau^{1-\alpha}} = \frac{f_k(k)}{(-ik)^\alpha}. \quad (6.82)
$$

### 6.4.2.2 *Fourier transform of the Riemann-Liouville fractional derivative and the Riesz fractional derivative*

The Riemann-Liouville fractional derivative in the interval $(-\infty, x)$ is given by

$$_{-\infty}D_x^\alpha f(x) = \frac{1}{\Gamma(n-\alpha)} \frac{d^n}{dx^n} \int_{-\infty}^x \frac{f(y)}{(x-y)^{\alpha+1-n}} dy, \ n-1 < \alpha < n. \ (6.83)$$

Note that the Riemann-Liouville fractional derivative may be written in terms of the Caputo fractional derivative ($_{-\infty}^C D_x^\alpha$) as follows. First, we transform the Riemann-Liouville fractional derivative $_aD_x^\alpha f(x)$ to

$$_aD_x^\alpha f(x) = \frac{1}{\Gamma(n-\alpha)} \frac{d^n}{dx^n} \int_a^x \frac{f(y)}{(x-y)^{\alpha+1-n}} dy$$

$$= \frac{1}{(n-\alpha)\Gamma(n-\alpha)} \frac{d^n}{dx^n} \int_0^{(x-a)^{n-\alpha}} f\left(x - u^{\frac{1}{n-\alpha}}\right) du, \quad (6.84)$$

with the variable $u$ given by

$$u = (x-y)^{n-\alpha}. \quad (6.85)$$

Using the Leibnitz rule for integrals,

$$\frac{d}{dt} \int_{a(t)}^{b(t)} f(x,t)dx$$

$$= f(b(t),t)\frac{db(t)}{dt} - f(a(t),t)\frac{da(t)}{dt} + \int_{a(t)}^{b(t)} \frac{\partial f(x,t)}{\partial t}dx, \quad (6.86)$$

we obtain

$$_aD_x^\alpha f(x) = \frac{1}{(n-\alpha)\Gamma(n-\alpha)} \int_0^{(x-a)^{n-\alpha}} \frac{d^n f\left(x - u^{\frac{1}{n-\alpha}}\right)}{d\left(x - u^{\frac{1}{n-\alpha}}\right)^n} du$$

$$+ \sum_{j=0}^{n-1} \frac{(x-a)^{j-\alpha} f^{(j)}(a)}{\Gamma(j-\alpha+1)}. \quad (6.87)$$

Transform the integral of Eq. (6.87) to the original variable $y$ yields

$$_aD_x^\alpha f(x) = \frac{1}{\Gamma(n-\alpha)} \frac{d^n}{dx^n} \int_a^x \frac{f(y)}{(x-y)^{\alpha+1-n}} dy$$

$$= \frac{1}{\Gamma(n-\alpha)} \int_a^x \frac{f^{(n)}(\tau)}{(x-\tau)^{\alpha+1-n}} d\tau + \sum_{j=0}^{n-1} \frac{(x-a)^{j-\alpha} f^{(j)}(a)}{\Gamma(j-\alpha+1)}$$

$$= {}_{a}^{C}D_{x}^{\alpha}f(x) + \sum_{j=0}^{n-1} \frac{(x-a)^{j-\alpha}f^{(j)}(a)}{\Gamma(j-\alpha+1)}, \quad n-1 < \alpha < n. \tag{6.88}$$

Considering that the function $f(x)$ and its derivatives decay to zero for $a \to -\infty$, we have, in this case, the Riemann-Liouville fractional derivative equivalent to the Caputo fractional derivative [25], i.e,

$$_{-\infty}D_{x}^{\alpha}f(x) = \; _{-\infty}^{C}D_{x}^{\alpha}f(x) = \frac{1}{\Gamma(n-\alpha)} \int_{-\infty}^{x} \frac{f^{(n)}(\tau)}{(x-\tau)^{\alpha+1-n}}d\tau. \tag{6.89}$$

Applying the Fourier transform to Eq. (6.89) and then using the Fubini's theorem and the result (6.79) we obtain

$$\mathcal{F}\left[_{-\infty}D_{x}^{\alpha}f(x)\right] = (ik)^{\alpha-n}\,\mathcal{F}\left[f^{(n)}(x)\right]$$

$$= (ik)^{\alpha-n}\,(ik)^{n}\,f_{k}(k) = (ik)^{\alpha}\,f_{k}(k)\,. \tag{6.90}$$

For the interval $(x, \infty)$ the procedure is similar to the previous case, and the result is given by

$$\mathcal{F}\left[_{x}D_{\infty}^{\alpha}f(x)\right] = (-ik)^{\alpha}\,f_{k}(k)\,. \tag{6.91}$$

Combining the two intervals we obtain the following result to the Fourier transform of the Riesz fractional derivative:

$$\mathcal{F}\left[\frac{d^{\alpha}f(x)}{d|x|^{\alpha}}\right] = \mathcal{F}\left[-\frac{1}{2\cos(\pi\alpha/2)}\left(_{-\infty}D_{x}^{\alpha}f(x) + _{x}D_{\infty}^{\alpha}f(x)\right)\right]$$

$$= -\frac{1}{2\cos(\pi\alpha/2)}\left[(ik)^{\alpha} + (-ik)^{\alpha}\right]f_{k}(k)$$

$$= -\frac{|k|^{\alpha}f_{k}(k)}{2\cos(\pi\alpha/2)}\left[e^{\frac{i\pi\alpha}{2}} + e^{-\frac{i\pi\alpha}{2}}\right] = -|k|^{\alpha}f_{k}(k). \tag{6.92}$$

## Bibliography

[1] E. W. Montroll and G. H. Weiss, *J. Math. Phys.* **6**, 167 (1965).

[2] R. Metzler and J. Klafter, *Phys. Rep.* **339**, 1 (2000).

[3] H. Scher and M. Lax, *Phys. Rev. B* **7**, 4502 (1973).

[4] H. Scher and E. Montroll, *Phys. Rev. B* **12**, 2455 (1975).

[5] M. F. Shlesinger, *J. Stat. Phys.* **10**, 421 (1974).

[6] J. Klafter and R. Silbey, *Phys. Rev. Lett* **44**, 55 (1980).

[7] M. Kotulski, *J. Stat. Phys.* **81**, 777 (1995).

[8] E. Barkai and J. Klafter, Phys. Rev. Lett **79**, 2245 (1997).

[9] W. T. Coffey, D. S. F. Crothers, D. Holland and S. V. Titov, *J. Mol. Liq.* **114**, 165 (2004).

[10] H. C. Fogedby, *Phys. Rev. Lett.* **73**, 2517 (1994).

[11] H. C. Fogedby, *Phys. Rev. E* **50**, 1657 (1994).

[12] J. Klafter, A. Blumen and M. F. Shlesinger, *Phys. Rev. A* **35**, 3081 (1987).

[13] B. Berkowitz, A. Cortis, M. Dentz and H. Scher, *Rev. Geophysics* **44**, RG2003 (2006).

[14] A. Compte, *Phys. Rev. E* **55**, 6821 (1997).

[15] A. S. Reddy, D. T. Warshaviak and M. Chachisvilis, *Bioch. et Biophys. Acta* **1818**, 2271 (2012).

[16] E. Scalas, R. Gorenflo and F. Mainardi, *Physica A* **284**, 376 (2000).

[17] F. Mainardi F, M. Raberto, R. Gorenflo and E. Scalas, *Physica A* **287**, 468 (2000).

[18] K. S. Fa, *Mod. Phys. Lett. B* **26**, 1250151 (2012).

[19] K. Kim and S. M. Yoon, *Fractals* **11**, 131 (2003).

[20] T. Kaizoji and M. Kaizoji, *Physica A* **336**, 563 (2004).

[21] E. Scalas, T. Kaizoji, M. Kirchler, J. Huber and A. Tedeschi, *Physica A* **366**, 463 (2006).

[22] K. S. Fa and K. G. Wang, *Phys. Rev. E* **81**, 051126 (2010).

[23] E. Barkai, R. Metzler and J. Klafter, *Phys. Rev. E* **61**, 132 (2000).

[24] R. Gorenflo and F. Mainardi, "Fractional calculus: Integral and differential equations of fractional order" in *Fractals and Fractional Calculus in Continuum Mechanics*, edited by A. Carpinteri and F. Mainardi (Springer, Wien, 1997), pp. 223-276.

[25] I. Podlubny, *Fractional Differential Equations* (Academic Press, USA, 1999).

[26] A. Compte, *Phys. Rev. E* **53**, 4191 (1996).

[27] R. Metzler, J. Klafter and I. Sokolov, *Phys. Rev. E* **58**, 1621 (1998).

[28] J. P. Bouchaud and A. Georges, *Phys. Rep.* **195**, 12 (1990).

[29] B. J. West, P. Grigolini, R. Metzler and T. F. Nonnenmacher, *Phys. Rev. E* **55**, 99 (1997).

[30] S. Jespersen, R. Metzler and H. C. Fogedby, *Phys. Rev. E* **59**, 2736 (1999).

[31] H. A. Kramers, *Physica* **7**, 284 (1940).

[32] H. Risken, *The Fokker-Planck Equation: Methods of Solution and Applications*, 2rd ed. (Springer, Berlin, 1996).

[33] R. Metzler and J. Klafter, *J. Phys. Chem. B* **104**, 3851 (2000).

[34] R. Metzler, *Phys. Rev. E* **62**, 6233 (2000).

[35] S. Chandrasekhar, *Rev. Mod. Phys.* **15**, 1 (1943).

[36] I. S. Gradshteyn and I. M. Ryzhik, *Table of Integrals, Series and Products* (Academic Press, USA, 1965).

[37] P. Allegrini, G. Aquino, P. Grigolini, L. Palatella and A. Rosa, *Phys. Rev. E* **68**, 056123 (2003).

[38] K. S. Fa and K. G. Wang, *Phys. Rev. E* **81**, 011126 (2010).

[39] G. M. Zaslavsky, *Phys. Rep.* **371**, 461 (2002).

[40] O. P. Agrawal, *J. Phys. A* **40**, 6287 (2007).

[41] S. G. Samko, A. A. Kilbas and O. I. Marichev, *Fractional Integrals and Derivatives* (Gordon and Breach Science, Singapore, 1993).

## Chapter 7

# Uncoupled Continuous Time Random Walk model and its solution

### 7.1 Introduction

In order to get deeper insights into the integro-differential equations, different quantities related to the models given in the previous chapter are investigated and analyzed. Methods for solving the integro-differential equations are also developed. In fact, analytical solutions for the PDF, first two moments, first passage time density, mean first passage time, the intermediate scattering function and the dynamic structure factor are obtained and investigated. To cover a variety of different behaviors for short, intermediate and long times we will take different waiting time PDFs. A generalized integro-differential Klein-Kramers equation is also presented for the description of the subdiffusive and superdiffusive regimes.

### 7.2 Integro-differential diffusion equation for force-free

#### 7.2.1 *The first two moments and PDF*

Solution of the integro-differential diffusion equation (6.28) can be obtained by using the Fourier-Laplace transform. The first two moments for the displacement can be obtained by using the following expression:

$$\frac{\mathrm{d}\langle x^n \rangle}{\mathrm{d}t} = \int_{-\infty}^{\infty} x^n \frac{\partial \rho(x,t)}{\partial t}\mathrm{d}x \ , \ n = 1,2 \ . \tag{7.1}$$

Substituting Eq. (6.28) into Eq. (7.1) yields

$$\frac{\mathrm{d}\langle x^n \rangle}{\mathrm{d}t} = \int_0^t g\left(t - t_1\right)\frac{\mathrm{d}\langle x^n \rangle}{\mathrm{d}t_1}\mathrm{d}t_1$$

$$+ C\frac{d}{dt}\int_0^t g\left(t - t_1\right)\int_{-\infty}^{\infty} x^n \frac{\partial^2 \rho(x,t)}{\partial x^2}\mathrm{d}x\mathrm{d}t_1 \ . \tag{7.2}$$

After integrating the second term of the right-hand side of Eq.(7.2) by parts twice, we obtain

$$\frac{d\langle x \rangle}{dt} = \int_0^t g(t-t_1) \frac{d\langle x \rangle}{dt_1} dt_1 \tag{7.3}$$

and

$$\frac{d\langle x^2 \rangle}{dt} = \int_0^t g(t-t_1) \frac{d\langle x^2 \rangle}{dt_1} dt_1$$

$$+ 2C \frac{d}{dt} \int_0^t g(t-t_1) \int_{-\infty}^{\infty} \rho(x,t) dx dt_1 \; , \tag{7.4}$$

where we also consider that $\lim_{x \to \pm\infty} \rho(x,t)$ decreases faster than $1/x$. The normalization of the PDF $\rho(x,t)$ requires that $\int_{-\infty}^{\infty} \rho(x,t) dx = 1$, then Eq.(7.4) can be rewritten as

$$\frac{d\langle x^2 \rangle}{dt} = \int_0^t g(t-t_1) \frac{d\langle x^2 \rangle}{dt_1} dt_1 + 2C \frac{d}{dt} \int_0^t g(t-t_1) dt_1 \; . \tag{7.5}$$

Applying the Laplace transform to Eqs. (7.3) and (7.5) we arrive at

$$\langle x \rangle_s = \frac{\langle x \rangle_0}{s} \tag{7.6}$$

and

$$\langle x^2 \rangle_s = \frac{\langle x^2 \rangle_0}{s} + \frac{2Cg(s)}{s\,[1 - g(s)]} \; , \tag{7.7}$$

where $\langle x \rangle_0 = \langle x(t=0) \rangle$ is the initial value for $\langle x \rangle$. Eqs. (7.6) and (7.7) are valid for generic waiting time PDF. Moreover, the first moment gives

$$\langle x \rangle = \langle x \rangle_0 \; , \tag{7.8}$$

for any waiting time PDF. Then, we set $\langle x \rangle = \langle x \rangle_0 = 0$.

To obtain exact solutions for the PDF $\rho(x,t)$ we first do Fourier inverse operation on $\rho_{ks}(k,s)$ in Eq.(6.22), and we have

$$\rho_s(x,s) = \frac{1 - g_s(s)}{2\pi C s g_s(s)} \int_{-\infty}^{\infty} \frac{e^{ikx}}{k^2 + \frac{1-g_s(s)}{Cg_s(s)}} dk \tag{7.9}$$

with $\rho_{k0}(k) = 1$. Performing the complex contour in Eq. (7.9), with $s \in R$, $(1 - g_s(s))/g_s(s) > 0$ and $s \neq 0$, we obtain

$$\rho_s(x,s) = \frac{1}{2\sqrt{C}s} \sqrt{\frac{1 - g_s(s)}{g_s(s)}} e^{-\frac{|x|}{\sqrt{C}} \sqrt{\frac{1-g_s(s)}{g_s(s)}}} \; . \tag{7.10}$$

We can also check the solution Eq. (7.10) by using a different calculation. By applying the Fourier transform to Eq. (7.10), we have

$$\rho_{ks}(k,s) = \frac{1}{2\sqrt{C}s}\sqrt{\frac{1-g_s(s)}{g_s(s)}} \int_{-\infty}^{\infty} e^{-ikx} e^{-\frac{|x|}{\sqrt{C}}\sqrt{\frac{1-g_s(s)}{g_s(s)}}} dx$$

$$= \frac{1}{2\sqrt{C}s}\sqrt{\frac{1-g_s(s)}{g_s(s)}} \left[ \int_{-\infty}^{0} e^{\left(-ik+\sqrt{\frac{1-g_s(s)}{Cg_s(s)}}\right)x} dx + \int_{0}^{\infty} e^{-\left(ik+\sqrt{\frac{1-g_s(s)}{Cg_s(s)}}\right)x} dx \right]$$

$$= \frac{1}{2\sqrt{C}s}\sqrt{\frac{1-g_s(s)}{g_s(s)}} \left[ \frac{1}{-ik+\sqrt{\frac{1-g_s(s)}{Cg_s(s)}}} + \frac{1}{ik+\sqrt{\frac{1-g_s(s)}{Cg_s(s)}}} \right]$$

$$= \frac{1-g_s(s)}{s\left[1-(1-Ck^2)g_s(s)\right]}. \tag{7.11}$$

Thus, we have recovered Eq. (6.22).

The $n$-th moment in Laplace space can be obtained from Eq. (7.10), and the result is given by

$$\langle x^n \rangle_s = \frac{n!}{s} \left[ \frac{Cg_s(s)}{1-g_s(s)} \right]^{\frac{n}{2}}, \tag{7.12}$$

where $n$ is an even number. For odd number the $n$-th moment is zero due to the fact that the PDF is symmetrical. Note that the second moment given by Eq. (7.12) is equal to (7.7). From Eq. (7.10) we can show the PDF is normalized, i.e,

$$\int_{-\infty}^{\infty} \rho(x,t)dx = \mathcal{L}^{-1} \int_{-\infty}^{\infty} \rho_s(x,s)dx = \mathcal{L}^{-1}\frac{1}{s} = 1 , \tag{7.13}$$

where $\mathcal{L}^{-1}$ denotes the inverse Laplace transform. Solutions for $\rho(x,t)$ can be obtained from Eq. (7.10) by using several methods for inverting the Laplace transform, and they can be given in different representations.

*Case 1.* The waiting time PDF is given by a combination of the power-law and generalized Mittag-Leffler function (6.36), i.e,

$$g_1(t) = -\frac{d}{dt}E_{\alpha,1}\left(-\lambda_\alpha t^\alpha\right) = \lambda_\alpha t^{\alpha-1}E_{\alpha,\alpha}(-\lambda_\alpha t^\alpha), 0 < \alpha \leq 1. \tag{7.14}$$

The waiting time PDF (7.14) has a power-law tail, and it has a divergent characteristic waiting time. The Laplace transform of $g_1(t)$ is given by

$$g_{s1}(s) = \frac{\lambda_\alpha}{\lambda_\alpha + s^\alpha} . \tag{7.15}$$

Note that the $g_1(t)$ is normalized; it is easy to show that $\int_0^\infty g_1(t)dx = 1$ or $\lim_{s\to 0} g_{s1}(s) = 1$. The solution for $\rho_1(x,t)$ can be obtained as follows. Substituting Eq. (7.15) into Eq. (7.10) and expanding the result in Taylor series yields

$$\rho_{s1}(x,s) = \frac{1}{\sqrt{4C\lambda_\alpha}} \sum_{n=0}^{\infty} \left(\frac{x^2}{C\lambda_\alpha}\right)^{\frac{n}{2}} \frac{(-1)^n}{n!s^{1-\frac{\alpha}{2}(1+n)}} . \tag{7.16}$$

Applying the Laplace inversion to $\rho_{s1}(x,s)$ we obtain the following solution:

$$\rho_1(x,t) = \frac{1}{\sqrt{4C\lambda_\alpha t^\alpha}} \sum_{n=0}^{\infty} \frac{(-1)^n}{n!\Gamma\left(1-\alpha\frac{1+n}{2}\right)} \left(\frac{x^2}{C\lambda_\alpha t^\alpha}\right)^{\frac{n}{2}} . \tag{7.17}$$

This last result is exactly the solution of the fractional diffusion equation (6.23) [1]. The solution (7.17) can also be written in terms of the Wright function:

$$\rho_1(x,t) = \frac{1}{\sqrt{4C\lambda_\alpha t^\alpha}} W_{-\frac{\alpha}{2},1-\frac{\alpha}{2}} \left(\frac{-|x|}{\sqrt{C\lambda_\alpha t^\alpha}}\right) , \tag{7.18}$$

where $W_{\mu,\nu}(z)$ is the Wright function [2,3] (see Appendix 7.8.1). For $\alpha = 1$, $g_1(t)$ reduces to the exponential form and the PDF reduces to the Gaussian shape

$$\rho_1(x,t) = \frac{1}{\sqrt{4\pi C\lambda t}} e^{-\frac{x^2}{4\lambda Ct}} , \tag{7.19}$$

as is expected.

Substituting Eq. (7.15) into Eq. (7.7) we obtain the following subdiffusive behavior:

$$\langle x^2 \rangle = \frac{2C\lambda_\alpha}{\Gamma(1+\alpha)} t^\alpha , \tag{7.20}$$

with $\langle x^2 \rangle_0 = 0$. This anomalous behavior is due to the fact that the waiting time PDF $g_1(t)$ has a power-law asymptotic behavior

$$g_1(t) \sim \frac{1}{t^{1+\alpha}} , \tag{7.21}$$

and the corresponding characteristic waiting time is divergent.

*Case 2.* The waiting time PDF is given by the Gamma distribution:

$$g_2(t) = b^\beta t^{\beta-1} \exp(-bt)/\Gamma(\beta) , \ b > 0, \ \beta > 0. \tag{7.22}$$

It interpolates approximately between the initial power-law form and exponential behavior in the long-time limit, and the parameter $b$ controls the power-law behavior. The waiting time PDF $g_2(t)$ has been used, for instance, to describe the dynamics of biological systems such as kinesin molecules [4] and other proteins. The Laplace transform of $g_2(t)$ is given by

$$g_{s2}(s) = \frac{b^\beta}{(b+s)^\beta} \, . \tag{7.23}$$

The PDF solution corresponding to $g_2(t)$ may be obtained in an integral form [5] and it is given by [6]

$$\rho_2(x,t) = \frac{1}{2\pi\sqrt{Cb^\beta}} \int_0^\infty du \Phi_2(u,x) \cos(ut + \theta_2(u,x)), \tag{7.24}$$

where

$$\theta_2(u,x) = \frac{\theta_{22}(u) - \pi}{2} - \frac{|x|}{\sqrt{Cb^\beta}} \sqrt{r_{22}(u)} \sin\left(\frac{\theta_{22}(u)}{2}\right), \tag{7.25}$$

$$\Phi_2(u,x) = \frac{\sqrt{r_{22}(u)}}{u} e^{-\frac{|x|}{\sqrt{Cb^\beta}} \sqrt{r_{22}(u)} \cos\left(\frac{\theta_{22}(u)}{2}\right)}, \tag{7.26}$$

$$r_{22}(u) = \sqrt{\left(r_{21}^\beta(u) \cos(\beta\theta_{21}(u)) - b^\beta\right)^2 + r_{21}^{2\beta}(u) \sin^2(\beta\theta_{21}(u))}, \tag{7.27}$$

$$\theta_{22}(u) = \arctan\left(\frac{r_{21}^\beta(u) \sin(\beta\theta_{21}(u))}{r_{21}^\beta(u) \cos(\beta\theta_{21}(u)) - b^\beta}\right), \quad r_{21}(u) = \sqrt{u^2 + b^2} \tag{7.28}$$

and

$$\theta_{21}(u) = \arctan\left(\frac{u}{b}\right). \tag{7.29}$$

The asymptotic expansion of $\rho(x,t)$ (for a given $x$ and $t \gg 1$) is given by

$$\rho_2(x,t) \sim \frac{1}{2}\sqrt{\frac{\beta}{\pi Cbt}} \, . \tag{7.30}$$

Eq. (7.30) shows that the PDF $\rho_2(x,t)$ presents a decay $1/\sqrt{t}$ similar to the one of the normal diffusion, and independent of the spatial coordinate. This is not surprised because the waiting time PDF (7.22) has a finite characteristic waiting time.

Substituting Eq. (7.23) into Eq. (7.7) and expanding the result in Taylor series we obtain

$$\langle x^2 \rangle_s = 2C \sum_{n=0}^\infty \frac{b^{\beta(1+n)}}{s(b+s)^{\beta(1+n)}}. \tag{7.31}$$

Applying the Laplace inversion to (7.31) yields

$$\langle x^2 \rangle = 2C \sum_{n=0}^{\infty} \int_0^t d\tau \frac{b^{\beta(1+n)} \tau^{\beta(1+n)-1} \exp(-b\tau)}{\Gamma(\beta(1+n))}$$

$$= 2C \sum_{n=0}^{\infty} \left[ 1 - \frac{\Gamma(\beta(1+n), bt)}{\Gamma(\beta(1+n))} \right]. \tag{7.32}$$

For $bt \ll 1$ the second moment yields

$$\langle x^2 \rangle \sim \frac{2Cb^{\beta}}{\Gamma(1+\beta)} t^{\beta}, \tag{7.33}$$

whereas for $bt \gg 1$ it is given by

$$\langle x^2 \rangle \sim \frac{2Cb}{\beta)} t. \tag{7.34}$$

Eqs. (7.33) and (7.34) show that the second moment initially follows an anomalous regime (subdiffusive process for $0 < \beta < 1$ and superdiffusive process for $\beta > 1$) and it follows a linear regime in the long-time limit. The linear regime given by Eq. (7.34) is due to the fact that the characteristic waiting time and the jump length variance are finite.

*Case 3.* The waiting time PDF is given by

$$g_3(t) = A_N \sum_{i=0}^{N} c_i \exp(-\gamma_i t), \quad \gamma_i > 0, \ A_N = 1 / \sum_{j=0}^{N} c_j / \gamma_j; \tag{7.35}$$

it contains multiple characteristic times and it may exhibit power-law behavior with logarithmic oscillation at the intermediate times and exponential behavior in the long-time limit. Moreover, the function $g_3(t)$ has been used for different purposes to describe such as relaxation friction and fluorescence decay in physical and biological complex systems [7–9]. The Laplace transform of $g_3(t)$ is given by

$$g_{s3}(s) = A_N \sum_{i=0}^{N} \frac{c_i}{\gamma_i + s}. \tag{7.36}$$

In order to show some interesting aspects of the CTRW model with the waiting time PDF (7.35) we consider some specific choices of $c_i$ and $\gamma_i$. Let us take $N = 1$ and $A_1 = 1$; in particular, this model can reproduce the mean square displacement of the Langevin equation for Brownian motion of a particle with an inertial term. The normalization factor implies that

$$c_0 = \gamma_0 \left( 1 - \frac{c_1}{\gamma_1} \right). \tag{7.37}$$

Substituting Eq. (7.36) into Eq. (7.7), with the conditions (7.37) and the sharp position $x_0$, we obtain

$$\langle (x - x_0)^2 \rangle = \frac{2C}{E_3^2} \left[ -(E_2 - E_1 E_3)\left(1 - e^{-E_3 t}\right) + E_2 E_3 t \right], \qquad (7.38)$$

where

$$E_1 = \gamma_0 + \left(1 - \frac{\gamma_0}{\gamma_1}\right) c_1, \qquad E_2 = \gamma_0 \gamma_1 \qquad (7.39)$$

and

$$E_3 = \gamma_1 - \left(1 - \frac{\gamma_0}{\gamma_1}\right) c_1 . \qquad (7.40)$$

Now we can compare the result (7.38) with the one of the Langevin equation for Brownian motion [10]. The system has a particle which starts at time $t = 0$ at $x = x_0$ with the velocity $v = v_0$, and it is described by the following equation:

$$\frac{dv(t)}{dt} + \gamma v(t) = L(t), \qquad (7.41)$$

where $L(t)$ is the Langevin force with zero mean $\langle L(t) \rangle = 0$ and correlation function given by $\langle L(t_1) L(t_2) \rangle = q\delta(t_1 - t_2)$. The mean square displacement of the Brownian motion of a particle is given by

$$\langle (x - x_0)^2 \rangle = \left( v_0^2 - \frac{q}{2\gamma} \right) \frac{(1 - e^{-\gamma t})^2}{\gamma^2} - \frac{q}{\gamma^3}\left(1 - e^{-\gamma t}\right) + \frac{q}{\gamma^2}t \qquad (7.42)$$

and the diffusion constant $D$ given by $D = q/(2\gamma^2)$.

In order to reproduce the result (7.42) from Eq. (7.38) we first consider the initial velocity distribution for the stationary state given by $\langle v_0^2 \rangle = q/(2\gamma)$. Then, from Eq. (7.42) we have

$$\langle (x - x_0)^2 \rangle = -\frac{q}{\gamma^3}\left(1 - e^{-\gamma t}\right) + \frac{q}{\gamma^2}t , \qquad (7.43)$$

and from Eqs. (7.38)-(7.40) we obtain

$$\gamma_0 = \frac{q}{C\gamma^2} \frac{1}{\left(1 \pm \sqrt{1 - \frac{2q}{C\gamma^3}}\right)} , \qquad \gamma_1 = \frac{\gamma}{2}\left(1 \pm \sqrt{1 - \frac{2q}{C\gamma^3}}\right), \qquad c_0 = -c_1 ,$$
$$(7.44)$$

$$E_1 = 0, \qquad E_2 = \frac{q}{2C\gamma} \quad \text{and} \quad E_3 = \gamma . \qquad (7.45)$$

These results link the Langevin approach to the integro-differential diffusion equation (6.28). In fact, in the Langevin equation (7.41) the parameters $\gamma$

and $q$ are related to the macroscopic and microscopic characteristic times, respectively. In the integro-differential approach (7.38) these quantities are related to the parameters $\gamma_0$ and $\gamma_1$ of the waiting time PDF. Notice that the mean square displacement (7.43) has the ballistic diffusion for short times, and it follows a linear regime in the long-time limit.

The PDF $\rho(x, t)$ for $N = 1$ is given by

$$\rho_3(x, t) = \frac{1}{2\pi\sqrt{C}} \int_0^\infty du \Phi_3\left(u, x\right) \cos\left(ut + \theta_3\left(u, x\right)\right) \qquad (7.46)$$

where

$$\theta_3\left(u, x\right) = \frac{\theta_{31}\left(u\right) - \theta_{32}\left(u\right) - \pi}{2} - \frac{|x|}{\sqrt{C}} \sqrt{\frac{r_{31}\left(u\right)}{r_{32}\left(u\right)}} \sin\left(\frac{\theta_{31}\left(u\right) - \theta_{32}\left(u\right)}{2}\right), \qquad (7.47)$$

$$\Phi_3(u, x) = \sqrt{\frac{r_{31}\left(u\right)}{r_{32}\left(u\right)}} \frac{\exp\left(-\frac{|x|}{\sqrt{C}} \sqrt{\frac{r_{31}(u)}{r_{32}(u)}} \cos\left(\frac{\theta_{31}(u) - \theta_{32}(u)}{2}\right)\right)}{u}, \qquad (7.48)$$

$$r_{32}\left(u\right) = \left[\left(\gamma_0\gamma_1\right)^2 + \left(u\overline{A}_1\right)^2\right]^{\frac{1}{2}}, \qquad \theta_{32}\left(u\right) = \arctan\left(\frac{u\overline{A}_1}{\gamma_0\gamma_1}\right), \qquad (7.49)$$

$$r_{31}\left(u\right) = \left[u^4 + \left(\gamma_0 + \gamma_1 - \overline{A}_1\right)^2\right]^{\frac{1}{2}}, \qquad (7.50)$$

$$\theta_{31}\left(u\right) = \arctan\left(-\frac{\gamma_0 + \gamma_1 - \overline{A}_1}{u^2}\right) \qquad (7.51)$$

and $\overline{A}_1 = (c_0 + c_1) A_1$.

Usually the starting point to the analysis of the CTRW model is to give the waiting time PDF, such as the cases (1-3). However, we can start the initial point to the analysis from the second moment when the diffusion processes are given; for both cases the solutions can be obtained. Let us consider that the second moment is described by a combination of the power-law and the generalized Mittag-Leffler function as follows:

$$\langle x^2 \rangle = 2C\lambda_{\beta-1}t^{\beta-1}E_{\alpha,\beta}(-\bar{\lambda}_\alpha t^\alpha). \qquad (7.52)$$

Note that the second moment should be positive; in this case, the values of the parameters of the generalized Mittag-Leffler function should be restricted, i.e., they are given by $0 < \alpha \leq 1$ and $\beta \geq \alpha$ due to the fact that

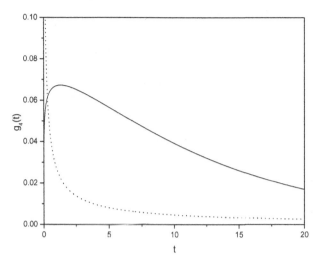

Fig. 7.1   Plots of $g_4(t)$. The dotted line corresponds to $\alpha = 0.3$, $\beta = 1.5$, $\lambda_{\beta-1} = 0.1$ and $\bar{\lambda}_\alpha = 0.5$, whereas the solid line corresponds to $\alpha = 0.3$, $\beta = 2.2$, $\lambda_{\beta-1} = 0.1$ and $\bar{\lambda}_\alpha = 0.5$

the generalized Mittag-Leffler function is completely monotonic, in these intervals [11]. This condition gives the generalized Mittag-Leffler function completely positive or negative. A simple numerical check can determine the positive values of the second moment. The $g(t)$ function can be obtained as follows. Applying the Laplace transform to Eq. (7.52) yields

$$\langle x^2 \rangle_s = \frac{2C\lambda_{\beta-1}s^{\alpha-\beta}}{\bar{\lambda}_\alpha + s^\alpha}. \tag{7.53}$$

Comparing Eq. (7.53) with Eq. (7.7) we obtain

$$g_{s4}(s) = \frac{\lambda_{\beta-1}s^{\alpha-\beta+1}}{\bar{\lambda}_\alpha + s^\alpha + \lambda_{\beta-1}s^{\alpha-\beta+1}}. \tag{7.54}$$

Eq. (7.54) is the $g(t)$ function in Laplace space which describes the second moment (7.52). The $g_1(t)$ function is recovered by setting $\bar{\lambda}_\alpha = 0$ and $\beta = 1 + \alpha$. Note that the $g_4(t)$ function is a PDF, therefore it should contain positive values. Then the parameter $\beta$ should be restricted to some interval which may be verified numerically by plotting the $g_4(t)$ function. The $g_4(t)$ function can be obtained from Eq. (7.54) by inverting the Laplace transform; a series representation of the solution can be obtained as follows. First we rewrite the expression (7.54) as

$$g_{s4}(s) = \frac{\lambda_{\beta-1}s^{\alpha-\beta+1}}{(\bar{\lambda}_\alpha + s^\alpha)\left(1 + \frac{\lambda_{\beta-1}s^{\alpha-\beta+1}}{\bar{\lambda}_\alpha + s^\alpha}\right)}. \tag{7.55}$$

Expanding Eq. (7.55) in series yields

$$g_{s4}(s) = -\sum_{k_1=0}^{\infty} \left[ \frac{-\lambda_{\beta-1}}{(\bar{\lambda}_\alpha + s^\alpha)s^{\beta-\alpha-1}} \right]^{1+k_1}$$

$$= -\sum_{k_1=0}^{\infty} \frac{(-\lambda_{\beta-1})^{k_1}}{(k_1-1)!} \sum_{k_2=0}^{\infty} \frac{(k_1+k_2-1)!(-\bar{\lambda}_\alpha)^{k_2}}{k_2! s^{k_1(\beta-1)+\alpha k_2}}. \tag{7.56}$$

Inverting the Laplace transform of Eq. (7.56) we obtain

$$g_4(t) = \lambda_{\beta-1} t^{\beta-2}$$

$$\times \sum_{k_1=0}^{\infty} \frac{(-\lambda_{\beta-1}t^{\beta-1})^{k_1}}{k_1!} \sum_{k_2=0}^{\infty} \frac{(k_1+k_2)! \left(-\bar{\lambda}_\alpha t^\alpha\right)^{k_2}}{k_2! \Gamma((\beta-1)(k_1+1)+\alpha k_2)}. \tag{7.57}$$

Besides, Eq. (7.57) can be written as (see Appendix (5.4.1))

$$g_4(t) = \lambda_{\beta-1} t^{\beta-2} \sum_{k_1=0}^{\infty} \frac{(-\lambda_{\beta-1}t^{\beta-1})^{k_1}}{k_1!} E_{\alpha,\beta-1+(\beta-\alpha-1)k_1}^{(k_1)} \left(-\bar{\lambda}_\alpha t^\alpha\right). \tag{7.58}$$

For $\bar{\lambda}_\alpha = 0$ and $\beta = 1+\alpha$, Eq. (7.57) reduces to

$$g_4(t) = \lambda_\alpha t^{\alpha-1} \sum_{n=0}^{\infty} \frac{(-\lambda_\alpha t^\alpha)^n}{\Gamma(\alpha+\alpha n)} = \lambda_\alpha t^{\alpha-1} E_{\alpha,\alpha}(-\lambda_\alpha t^\alpha), \tag{7.59}$$

which recovers the $g_1(t)$ function. The asymptotic behaviors of $g_4(t)$ are also obtained from Eq. (7.54); for short times it is described by

$$g_4(t) \sim \frac{\lambda_{\beta-1} t^{\beta-2}}{\Gamma(\beta-1)}, \tag{7.60}$$

whereas for long times it is given by

$$g_4(t) \sim -\frac{\bar{\lambda}_\alpha t^{\alpha-\beta}}{\lambda_{\beta-1}\Gamma(1+\alpha-\beta)}. \tag{7.61}$$

Eq. (7.60) shows that $g_4(t)$ is divergent at $t = 0$ for $\beta < 2$, whereas $g_4(t) \to 0$ for $\beta > 2$ and $t \to 0$. For large times, Eq. (7.61) shows a power-law decay for $g_4(t)$. Fig. 7.1 shows the behaviors of the $g_4(t)$ function and its asymptotic limits. With the intervals $0 < \alpha \le 1$ and $1+\alpha < \beta < 2+\alpha$, one can determine the asymptotic behaviors of the second moment (7.52). For short times it is given by

$$\langle x^2 \rangle \sim 2C\lambda_{\beta-1} t^{\beta-1}, \tag{7.62}$$

whereas for long times it is given by

$$\langle x^2 \rangle \sim \frac{2C\lambda_{\beta-1}}{\lambda_\alpha \Gamma(\beta-\alpha)} t^{\beta-(1+\alpha)}. \tag{7.63}$$

Eqs. (7.62)-(7.63) show that the model described by $g_4(t)$ can describe subdiffusion and superdiffusion for short times, and subdiffusion in the long-time limit.

### 7.2.2 First passage time density and mean first passage time

In this subsection we consider the first passage time whose dynamics of a particle diffusing in an interval $[-a_1, a_2]$ is governed by the integro-differential diffusion equation (6.28), and it is subject to absorbing boundaries, i.e.,

$$\rho(-a_1, t) = \rho(a_2, t) = 0 \tag{7.64}$$

and the initial condition given by $\rho(x, 0) = \delta(x - x_0)$.

An exact solution for $\rho(x, t)$ can be obtained from Eq. (6.28) by using the method of separation of variables, i.e,

$$\rho(x, t) = \rho_x(x)\rho_t(t) . \tag{7.65}$$

Substituting Eq. (7.65) into Eq. (6.28) yields

$$\frac{d^2 \rho_x(x)}{dx^2} = -\frac{\mu_n}{C}\rho_x(x) \tag{7.66}$$

and

$$\frac{d\rho_t(t)}{dt} = -\mu_n \frac{\partial}{\partial t} \int_0^t g^*(t - t_1) \rho_t(t_1) dt_1 , \tag{7.67}$$

where $\mu_n$ is the separation constant. The solution of Eq. (7.66), with the boundaries (7.64), is given by

$$\rho_{xn}(x) = A_n \sin\left(\frac{n\pi(x + a_1)}{a_1 + a_2}\right) , \tag{7.68}$$

with

$$\mu_n = C\left(\frac{n\pi}{a_1 + a_2}\right)^2 . \tag{7.69}$$

The solution for $\rho(x, t)$ is given by a sum of the solutions for space and time parts, i.e,

$$\rho(x, t) = \sum_{n=1}^{\infty} A_n \sin\left(\frac{n\pi(x + a_1)}{a_1 + a_2}\right) \rho_{tn}(t) . \tag{7.70}$$

The coefficients $A_n$ are determined by imposing the initial condition $\rho(x, 0) = \delta(x - x_0)$, which are given by

$$A_n = \frac{2}{a_1 + a_2} \sin\left(\frac{n\pi(x_0 + a_1)}{a_1 + a_2}\right) . \tag{7.71}$$

Substituting Eq. (7.71) into Eq. (7.70) yields

$$\rho(x,t) = \frac{2}{a_1 + a_2} \sum_{n=1}^{\infty} \sin\left(\frac{n\pi(x_0 + a_1)}{a_1 + a_2}\right) \sin\left(\frac{n\pi(x + a_1)}{a_1 + a_2}\right) \rho_{tn}(t) . \quad (7.72)$$

The FPT density function 3.6.1 is expressed in terms of $\rho(x,t)$ as

$$F(t) = -\frac{dI(t)}{dt} , \quad (7.73)$$

where

$$I(t) = \int_{-a_1}^{a_2} \rho(x,t)dx. \quad (7.74)$$

Substituting Eq. (7.72) into Eq. (7.73) we obtain

$$F(t) = -\frac{4}{\pi} \sum_{n=0}^{\infty} \frac{\sin\left(\frac{(1+2n)\pi(x_0+a_1)}{a_1+a_2}\right)}{(1+2n)} \frac{d\rho_{t(1+2n)}(t)}{dt} . \quad (7.75)$$

The mean first passage time is given by

$$M = \int_0^{\infty} \int_{-a_1}^{a_2} \rho(x,t)dxdt = \int_0^{\infty} I(t)dt . \quad (7.76)$$

Substituting Eq. (7.72) into Eq. (7.76) yields

$$M = \frac{4}{\pi} \sum_{n=0}^{\infty} \frac{\sin\left(\frac{(1+2n)\pi(x_0+a_1)}{a_1+a_2}\right)}{(1+2n)} \int_0^{\infty} \rho_{t(1+2n)}(t)dt . \quad (7.77)$$

In particular, for $a_1 = 0$, $a_2 = L$ and $x_0 = L/2$, Eqs. (7.72), (7.75) and (7.77) reduce to

$$\rho(x,t) = \frac{2}{L} \sum_{n=0}^{\infty} (-1)^n \sin\left(\frac{\pi(1+2n)x}{L}\right) \rho_{t(1+2n)}(t) , \quad (7.78)$$

$$F(t) = -\frac{4}{\pi} \sum_{n=0}^{\infty} \frac{(-1)^n}{(1+2n)} \frac{d\rho_{t(1+2n)}(t)}{dt} \quad (7.79)$$

and

$$M = \frac{4}{\pi} \sum_{n=0}^{\infty} \frac{(-1)^n}{(1+2n)} \int_0^{\infty} \rho_{t(1+2n)}(t)dt . \quad (7.80)$$

The explicit solution for $\rho_t(t)$ is obtained from Eq. (7.67) for a given $g(t)$. Applying the Laplace transform to Eq. (7.67) yields

$$\rho_{ts}(s) = \frac{1}{s\left[1 + \mu_n g_s^*(s)\right]} . \quad (7.81)$$

Now one considers the $g(t)$ function given in the previous section:

*Case a.* For $g_1(t)$ (Eq. (7.14)) , the solutions for $\rho(x,t)$ and $F(t)$ are given by

$$\rho(x,t) = \frac{2}{L} \sum_{n=0}^{\infty} (-1)^n \sin\left(\frac{\pi(1+2n)x}{L}\right) E_{\alpha,1}(-\mu_{1+2n}\lambda_\alpha t^\alpha) \qquad (7.82)$$

and

$$F(t) = \frac{4\lambda_\alpha t^{\alpha-1}}{\pi} \sum_{n=0}^{\infty} \frac{(-1)^n \mu_{1+2n}}{(1+2n)} E_{\alpha,\alpha}(-\mu_{1+2n}\lambda_\alpha t^\alpha) . \qquad (7.83)$$

It can be demonstrated that the MFPT related to $g_1(t)$ is divergent due to the long tail of the generalized Mittag-Leffler function [12, 13] The asymptotic limit of $\rho_{t(n)}(t)$ is given by

$$\rho_{t(n)}(t) \sim \frac{t^{-\alpha}}{\lambda_\alpha \mu_n \Gamma(1-\alpha)}. \qquad (7.84)$$

Substituting Eq. (7.84) into Eq. (7.80) yields

$$M \sim \frac{L^2}{8\lambda_\alpha C \Gamma(1-\alpha)} \int_{0}^{\infty} t^{-\alpha}dt . \qquad (7.85)$$

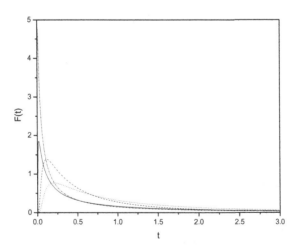

Fig. 7.2 Plots of the FPT density function (7.83). The parameters $C$ and $L$ have the values $C = 0.2$ and $L = 1$. The solid line corresponds to $\alpha = 0.5$ and $\lambda_\alpha = 0.5$. The dotted line corresponds to $\alpha = 0.8$ and $\lambda_\alpha = 0.5$. The dash-dotted line corresponds to $\alpha = 0.5$ and $\lambda_\alpha = 0.8$. The dashed line corresponds to $\alpha = 0.8$ and $\lambda_\alpha = 0.8$.

Thus, the integral (7.85) is divergent for any $\alpha < 1$. For $\alpha = 1$ the above results ( Eqs. (7.82) and (7.83) ) reduce to those of the ordinary diffusion equation, and the Mittag-Leffler function decays like the exponential function; and the MFPT is given by

$$M_{ord.} = \frac{L^2}{8\lambda C}.  \tag{7.86}$$

Fig. 7.2 shows the FPT distribution (7.83) for $L = 1$ and $C = 0.2$. The peaks are more pronounced for $\alpha = 0.5$ than those obtained from $\alpha = 0.8$.

*Case b.* In the case of $g_2(t)$ (Eq. (7.22)), a solution for $\rho(x, t)$ can be obtained as follows. Expanding (7.81) in series using binomial series yields

$$\rho_{ts}(s) = \frac{1 - g_s(s)}{s\left[1 - (1 - \mu_n)\, g_s(s)\right]} = \frac{1 - g_s(s)}{s\,(1 + \mu_n)\left[1 - \left(\frac{\mu_n}{1+\mu_n} + \frac{1-\mu_n}{1+\mu_n}\, g_s(s)\right)\right]}$$

$$= \frac{1 - g_s(s)}{s\,(1 + \mu_n)} \sum_{k_1=0}^{\infty} \left[\frac{\mu_n}{1+\mu_n} + \frac{1-\mu_n}{1+\mu_n}\, g_s(s)\right]^{k_1} = \frac{1}{1+\mu_n}$$

$$\times \sum_{k_1=0}^{\infty} \left(\frac{\mu_n}{1+\mu_n}\right)^{k_1} \sum_{k_2=0}^{k_1} \binom{k_1}{k_2} \left(\frac{1-\mu_n}{\mu_n}\right)^{k_2} \frac{g_s^{k_2}(s) - g_s^{1+k_2}(s)}{s}. \tag{7.87}$$

Substituting Eq. (7.23) into Eq. (7.87) and then inverting the Laplace transform yields

$$\rho(x, t) = \frac{2}{L} \sum_{n=0}^{\infty} \frac{(-1)^n \sin\left(\frac{\pi(1+2n)x}{L}\right)}{1 + \mu_{1+2n}} \sum_{k_1=0}^{\infty} \left(\frac{\mu_{1+2n}}{1+\mu_{1+2n}}\right)^{k_1} \sum_{k_2=0}^{k_1} \binom{k_1}{k_2}$$

$$\times \left(\frac{1-\mu_{1+2n}}{\mu_{1+2n}}\right)^{k_2} \left[\frac{\Gamma\left(\beta\left(1+k_2\right), bt\right)}{\Gamma\left(\beta\left(1+k_2\right)\right)} - \frac{\Gamma\left(\beta k_2, bt\right)}{\Gamma\left(\beta k_2\right)}\right]. \tag{7.88}$$

For $F(t)$ and $M$ they are given by

$$F(t) = -\frac{4}{\pi} \sum_{n=0}^{\infty} \frac{(-1)^n}{(1+2n)(1+\mu_{1+2n})} \sum_{k_1=0}^{\infty} \left(\frac{\mu_{1+2n}}{1+\mu_{1+2n}}\right)^{k_1} \sum_{k_2=0}^{k_1} \binom{k_1}{k_2}$$

$$\times \left(\frac{1-\mu_{1+2n}}{\mu_{1+2n}}\right)^{k_2} b^{\beta k_2} \left[\frac{1}{\Gamma(\beta k_2)} - \frac{(bt)^\beta}{\Gamma(\beta(1+k_2))}\right] t^{\beta k_2 - 1} e^{-bt} \tag{7.89}$$

and

$$M = \frac{\beta L^2}{8bC}. \tag{7.90}$$

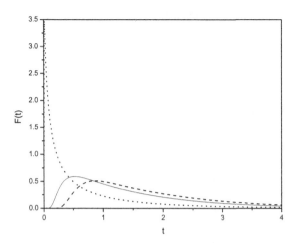

Fig. 7.3 Plots of the FPT density function (7.89). The parameters $b$, $C$ and $L$ have the values $b = 0.4$, $C = 0.2$ and $L = 1$. The dotted line corresponds to $\beta = 0.5$. The solid line corresponds to $\beta = 1$. The dashed line corresponds to $\beta = 1.2$.

One can see that the MFPT (7.90) gives a simple relation and it is proportional to $M_{ord.}$,

$$M = \beta M_{ord.},\qquad(7.91)$$

with $b = \lambda$. Fig. 7.3 shows the FPT distribution (7.89) for $L = 1$ and $C = 0.2$.

*Case c.* For $g_4(t)$ (Eq. (7.57)), the solutions for $\rho(x,t)$ and $F(t)$ are given by

$$\rho(x,t) = \frac{2}{L}$$

$$\times \sum_{n=0}^{\infty} \frac{(-1)^n \sin\left(\frac{\pi(1+2n)x}{L}\right)}{1 + \mu_{1+2n}} \sum_{k_1=0}^{\infty} \left(\frac{\mu_{1+2n}}{1 + \mu_{1+2n}}\right)^{k_1} \sum_{k_2=0}^{k_1} \binom{k_1}{k_2} \left(-\lambda_{\beta-1}t^{\beta-1}\right)^{k_2}$$

$$\times \left[\frac{1}{\Gamma\left(1 + (\beta-1)k_2\right)} + k_2 \sum_{k_3=1}^{\infty} \frac{(k_2 + k_3 - 1)!\left(-\bar{\lambda}_\alpha t^\alpha\right)^{k_3}}{k_2! k_3! \Gamma\left(1 + (\beta-1)k_2 + \alpha k_3\right)}\right] \quad (7.92)$$

and

$$F(t) = -\frac{4}{\pi}$$

$$\times \sum_{n=0}^{\infty} \frac{(-1)^n}{(1+2n)(1+\mu_{1+2n})} \sum_{k_1=1}^{\infty} \left(\frac{\mu_{1+2n}}{1+\mu_{1+2n}}\right)^{k_1} \sum_{k_2=1}^{k_1} \binom{k_1}{k_2} \frac{k_2}{t} \left(-\lambda_{\beta-1}t^{\beta-1}\right)^{k_2}$$

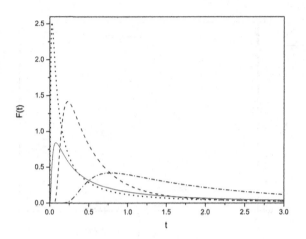

Fig. 7.4   Plots of the FPT density function (7.93) for $\bar{\lambda}_\alpha = \lambda^\alpha$ and $\lambda_{\beta-1} = \lambda^{\beta-1}$. The parameters $C$ and $L$ have the values $C = 0.2$ and $L = 1$. The solid line corresponds to $\alpha = 0.5$, $\beta = 1.7$ and $\lambda = 0.5$. The dotted line corresponds to $\alpha = 0.5$, $\beta = 1.7$ and $\lambda = 1.5$. The dashed line corresponds to $\alpha = 0.7$, $\beta = 2.2$ and $\lambda = 1.5$. The dash-dotted line corresponds to $\alpha = 0.5$, $\beta = 2.2$ and $\lambda = 0.5$.

$$\times \left[ \frac{\beta - 1}{\Gamma\left(1 + (\beta - 1)k_2\right)} + \sum_{k_3=1}^{\infty} \frac{(k_2 + k_3 - 1)!\left(-\bar{\lambda}_\alpha t^\alpha\right)^{k_3}}{k_2! k_3! \Gamma\left((\beta - 1)k_2 + \alpha k_3\right)} \right]. \tag{7.93}$$

The MFPT related to $g_4(t)$ is also divergent. The demonstration is similar to the previous case. Fig. 7.4 shows the FPT distribution (7.93) for $L = 1$ and $C = 0.2$. Figs. 7.2-7.4 show that the peak of the FPT distribution presents different height and time point.

### 7.2.3   Full decoupled case

For the full decoupled case without the presence of external force, a solution for the PDF $\rho(x,t)$ can be obtained from Eq. (6.17) which may be written as follows:

$$\rho_{ks}(k,s) = \Psi_s(s) \sum_{n=0}^{\infty} \left(g_s(s)\lambda_k(k)\right)^n. \tag{7.94}$$

The solution for $\rho(x,t)$ can be expressed in terms of the convolutions of the jump length and waiting time PDFs by inverting the Fourier and Laplace transforms of equation (7.94) [14]

$$\rho(x,t) = \sum_{n=0}^{\infty} g_n(t)\lambda_n(x) , \tag{7.95}$$

where

$$\lambda_n(x) = \int_{-\infty}^{\infty} \cdots \int_{-\infty}^{\infty} d\xi_{n-1} d\xi_{n-2} \cdots d\xi_1$$

$$\times \sigma\left(x - \xi_{n-1}\right) \sigma\left(\xi_{n-1} - \xi_{n-2}\right) \cdots \sigma\left(\xi_1\right) \qquad (7.96)$$

and

$$\times g_n(t) = \int_0^t \int_0^{\tau_1} \cdots \int_0^{\tau_{n-1}} d\tau_1 \cdots d\tau_n$$

$$g\left(t - \tau_1\right) g\left(\tau_1 - \tau_2\right) \cdots g\left(\tau_{n-1} - \tau_n\right) \Psi\left(\tau_n\right). \qquad (7.97)$$

However, the solution (7.95) is not convenient to be used except for the cases when $g_n(t)$ and $\lambda_n(x)$ have rapid convergence or their integrals can be evaluated analytically.

An alternative way to obtain solutions of Eq. (6.17) is to invert the Fourier transform directly, for a given $\lambda(x)$. Then, invert the Laplace transform for a given $g(t)$. To do so, let us consider the exponential jump length PDF given by

$$\lambda(x) = \frac{c}{2} \exp\left(-c\,|x|\right), \qquad (7.98)$$

where $c$ is a positive real number. It should be noted that the exponential jump length PDF gives a finite jump length variance. The Fourier transform of Eq. (7.98) yields

$$\lambda_k(k) = \frac{c^2}{c^2 + k^2}. \qquad (7.99)$$

Before calculating the inverse Fourier transform of Eq. (6.17) we rewrite it in the following form:

$$\rho_{ks}(k, s) = \frac{(1 - g_s(s))}{s} \left[1 + \frac{g_s(s)\lambda_k(k)}{1 - g_s(s)\lambda_k(k)}\right]. \qquad (7.100)$$

Substituting Eq. (7.99) into Eq. (7.100) and performing the complex contour integral, with $1 - g_s(s) > 0$ and $s \neq 0$, we obtain

$$\rho_s(x, s) = \frac{(1 - g_s(s))}{s} \left[\delta(x) + \frac{cg_s(s) \exp\left(-c\sqrt{1 - g_s(s)}\,|x|\right)}{2\sqrt{1 - g_s(s)}}\right]. \qquad (7.101)$$

We can show the PDF (7.101) is normalized, i.e,

$$\int_{-\infty}^{\infty} \rho(x, t)dx = \mathcal{L}^{-1} \int_{-\infty}^{\infty} \rho_s(x, s)dx = \mathcal{L}^{-1}\frac{1}{s} = 1. \qquad (7.102)$$

Moreover, we can obtain the $n$-moment (even $n$) in Laplace space from Eq. (7.101), which is given by

$$\langle x^n \rangle_s = \frac{n! g_s(s)}{c^n s \left(1 - g_s(s)\right)^{\frac{n}{2}}} . \qquad (7.103)$$

In particular, the second moment is given by

$$\langle x^2 \rangle_s = \frac{2 g_s(s)}{c^2 s \left(1 - g_s(s)\right)} . \qquad (7.104)$$

Note that the second moment (7.104) has the same result as the one given by the integro-differential diffusion equation (7.7) for any waiting time PDF, with $C = 1/c^2$. This means that the model described by (6.28) can preserve some characteristics of the full decoupled model.

In the case of $g_1(t)$ (Eq. (7.14) ) the explicit solution for the PDF is given by

$$\rho_1(x,t) = E_{\alpha,1} \left(-\lambda_\alpha t^\alpha\right) \delta(x)$$

$$+ \frac{c \lambda_\alpha t^\alpha}{2} \sum_{n=0}^{\infty} \frac{(-c|x|)^n}{n!(n+1)!!} \sum_{m=0}^{\infty} \frac{(2m+n+1)!!}{m! \Gamma(1+\alpha+\alpha m)} \left(-\frac{\lambda_\alpha t^\alpha}{2}\right)^m . \qquad (7.105)$$

For $\alpha = 1$, $g_1(t)$ reduces to the exponential function, and the PDF (7.105) can be written as

$$\rho_1(x,t) = \exp(-\lambda t) \delta(x)$$

$$+ \frac{c \lambda t}{2} \sum_{n=0}^{\infty} \frac{(-c|x|)^n}{n!} \, _1F_1 \left(\frac{3+n}{2}, 2, -\lambda t\right), \qquad (7.106)$$

where $_1F_1(a, b, z)$ is the Kummer confluent hypergeometric function [15]. Eq. (7.105) shows a sharp peak at the origin due to the Dirac delta function.

The second moment, related to $g_1(t)$, gives subdiffusive behavior for all the times

$$\langle x^2 \rangle = \frac{2\lambda_\alpha}{c^2 \Gamma(1+\alpha)} t^\alpha. \qquad (7.107)$$

Fig. 7.5 shows the PDF (7.105) without the first term and it is compared to the PDF (7.17) for large times, which are close to each other. The numerical results of PDFs may be obtained from Eqs. (7.17) and (7.105) or from Eqs. (7.10) and (7.101) by using a numerical inversion of Laplace transform algorithm [16]. The algorithm for Laplace inversion is especially useful when the analytical solution is difficult to be computed numerically.

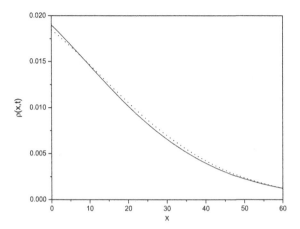

Fig. 7.5 Plots of the PDF versus position with $c = 0.2$, $\lambda_\alpha = 0.5$, $\alpha = 0.8$ and $t = 60$. The dotted line is described by Eq. (7.17), whereas the solid line is described by Eq. (7.105) without the first term.

For the waiting time PDF $g_2(t)$ the solution can be obtained in an integral form as

$$\rho_2(x, t) = \int_t^\infty du g_2(u) \delta(x)$$

$$+ \frac{cb^\beta}{2\pi} \int_0^\infty du \Phi_2(u, x) \cos(ut + \theta_2(u, x)), \quad (7.108)$$

where

$$\theta_2(u, x) = \frac{\theta_{22}(u) - \pi - 3\beta\theta_{21}(u)}{2}$$

$$- c\sqrt{\frac{r_{22}(u)}{r_{21}^\beta(u)}} \sin\left(\frac{\theta_{22}(u) - \beta\theta_{21}(u)}{2}\right) |x|, \quad (7.109)$$

$$\Phi_2(u, x) = \sqrt{\frac{r_{22}(u)}{r_{21}^{3\beta}(u)}} \frac{e^{-c\sqrt{\frac{r_{22}(u)}{r_{21}^\beta(u)}} \cos\left(\frac{\theta_{22}(u) - \beta\theta_{21}(u)}{2}\right) |x|}}{u}, \quad (7.110)$$

$$r_{22}(u) = \sqrt{r_{22x}^2(u) + r_{22y}^2(u)}, \quad \theta_{22}(u) = \arctan\left(\frac{r_{22y}(u)}{r_{22x}(u)}\right), \quad (7.111)$$

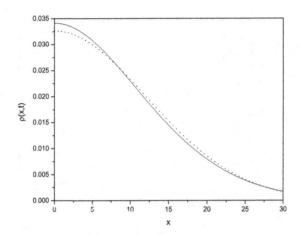

Fig. 7.6   Plots of the PDF versus position with $c = 0.5$, $b = 0.1$, $\beta = 0.8$ and $t = 150$. The dotted line is described by Eq. (7.24), whereas the solid line is described by Eq. (7.108) without the first term.

$$r_{22x}\left(u\right) = r_{21}^{\beta}\left(u\right)\cos\left(\beta\theta_{21}\left(u\right)\right) - b^{\beta},\ r_{22y}\left(u\right) = r_{21}^{\beta}\left(u\right)\sin\left(\beta\theta_{21}\left(u\right)\right),$$
(7.112)

$$r_{21}\left(u\right) = \sqrt{u^2 + b^2}\ \text{ and }\ \theta_{21}\left(u\right) = \arctan\left(\frac{u}{b}\right).$$
(7.113)

The second moment is given by

$$\left\langle x^2\right\rangle = \frac{2}{c^2}\left[-1 + \exp\left(-bt\right)E_{\beta,1}\left((bt)^{\beta}\right)\right.$$

$$\left. + b\int_0^t du \exp\left(-bu\right)E_{\beta,1}\left((bu)^{\beta}\right)\right].$$
(7.114)

Fig. 7.6 shows the PDF $\rho_2(x,t)$ versus position without the first term of equation (7.108) and the PDF (7.24), for large time.

For the waiting time PDF $g_3(t)$ ((7.35)) with $N = 1$, the solution can be obtained in an integral form as

$$\rho_3(x,t) = \int_t^{\infty} du g_3(u)\delta\left(x\right)$$

$$+ \frac{cA_1}{2\pi}\int_0^{\infty} du \Phi_3\left(u, x\right)\cos\left(ut + \theta_3\left(u, x\right)\right),$$
(7.115)

where

$$\theta_3\left(u, x\right) = \frac{2\theta_{33}\left(u\right) + \theta_{34}\left(u\right) - \pi}{2} - c\sqrt{r_{34}\left(u\right)}\sin\left(\frac{\theta_{34}\left(u\right)}{2}\right)|x|, \quad (7.116)$$

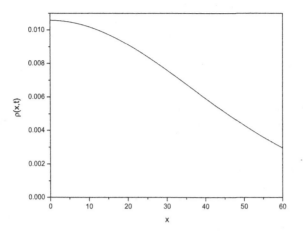

Fig. 7.7 The PDF versus position with $c = 0.2$, $c_0 = 0.5$ and $c_1 = 0.5$, $\gamma_0 = 0.5$, $\gamma_1 = 0.5$ and $t = 60$. The dotted line is described by Eq. (7.46), whereas the solid line is described by (7.115) without the first term.

$$\Phi_3\left(u, x\right) = \frac{r_{33}\left(u\right)\sqrt{r_{34}\left(u\right)}}{u}e^{-c\sqrt{r_{34}(u)}\cos\left(\frac{\theta_{34}(u)}{2}\right)|x|}, \tag{7.117}$$

$$r_{34}\left(u\right) = \sqrt{\left(1 - A_1 r_{33x}\left(u\right)\right)^2 + A_1^2 r_{33y}^2\left(u\right)}, \tag{7.118}$$

$$r_{33}\left(u\right) = \sqrt{r_{33x}^2\left(u\right) + r_{33y}^2\left(u\right)}, \tag{7.119}$$

$$\theta_{34}\left(u\right) = \arctan\left(\frac{A_1 r_{33y}\left(u\right)}{1 - A_1 r_{33x}\left(u\right)}\right), \; \theta_{33}\left(u\right) = \arctan\left(-\frac{r_{33y}\left(u\right)}{r_{33x}\left(u\right)}\right), \tag{7.120}$$

$$r_{33x}\left(u\right) = \frac{c_0\cos\left(\theta_{31}\left(u\right)\right)}{r_{31}\left(u\right)} + \frac{c_1\cos\left(\theta_{32}\left(u\right)\right)}{r_{32}\left(u\right)}, \tag{7.121}$$

$$r_{33y}\left(u\right) = \frac{c_0\sin\left(\theta_{31}\left(u\right)\right)}{r_{31}\left(u\right)} + \frac{c_1\sin\left(\theta_{32}\left(u\right)\right)}{r_{32}\left(u\right)}, \; \theta_{32}\left(u\right) = \arctan\left(\frac{u}{\gamma_1}\right), \tag{7.122}$$

$$r_{32}\left(u\right) = \sqrt{u^2 + \gamma_1^2}, \; r_{31}\left(u\right) = \sqrt{u^2 + \gamma_0^2} \; \text{and} \; \theta_{31}\left(u\right) = \arctan\left(\frac{u}{\gamma_0}\right). \tag{7.123}$$

The second moment is given by

$$\langle x^2 \rangle = \frac{2\gamma_0\gamma_1}{c^2}$$

$$\times \left\{ \frac{c_0\gamma_1 + c_1\gamma_0}{c_0\gamma_1^2 + c_1\gamma_0^2} t + \frac{c_0 c_1 (\gamma_0 - \gamma_1)^2}{(c_0\gamma_1^2 + c_1\gamma_0^2)^2} \left[ 1 - e^{-\frac{c_0\gamma_1^2 + c_1\gamma_0^2}{c_0\gamma_1 + c_1\gamma_0} t} \right] \right\}. \qquad (7.124)$$

The last expression shows the normal diffusion for $t \ll 1$ and $t \gg 1$. Fig. 7.7 shows the PDF $\rho_3(x, t)$ without the first term of equation (7.115) versus position and the PDF (7.46), for large time.

Details of the PDF in the full decoupled case, with the waiting time PDFs given above, have been described in Ref. [17], and the following characteristics of the PDF have been identified:

i) The PDF shows cusps at the origin for small times, even the second moment demonstrates normal diffusion.

ii) The PDF shows smooth shapes at large times when the second moment demonstrates normal diffusion.

iii) The peak of the PDF gradually increases to reaching a maximum with the time, and then decreasing with the time increase. This shows that the walker does not spread out for small times (in statistical sense).

A second jump length PDF considered here is the Gaussian function

$$\lambda(x) = \frac{1}{\sqrt{4\pi\sigma^2}} \exp\left( -\frac{x^2}{4\sigma^2} \right), \qquad (7.125)$$

where $\sigma$ is a real number. The Gaussian jump length PDF also gives a finite jump length variance. The Fourier transform of Eq. (7.125) is given by

$$\lambda_k(k) = \exp\left( -\sigma^2 k^2 \right). \qquad (7.126)$$

The Fourier inversion of Eq. (7.100), with $\lambda_k(k)$ described by Eq. (7.126), is obtained by performing the complex contour integral for $q_s(s) > 0$ and $s \neq 0$. The contour in the complex plane used to evaluate the second term of Eq. (7.100) is shown in Fig. 7.8. For $x > 0$, the appropriate contour is one that closes in the upper half-plane, whereas, for $x < 0$, the appropriate contour is one that closes in the lower half-plane. In fact, the poles are outside the contour. The result is given by

$$\rho_s(x, s) = G_s(s) \left[ \delta(x) + \frac{g_s(s)}{\pi} \int_0^\infty dr \right.$$

$$\times \left. \frac{\exp\left( -\frac{|x|r}{\sqrt{2}} \right) \left[ \cos\left( \frac{\pi}{4} - \sigma^2 r^2 + \frac{|x|r}{\sqrt{2}} \right) - g_s(s) \cos\left( \frac{\pi}{4} + \frac{|x|r}{\sqrt{2}} \right) \right]}{1 - 2g_s(s)\cos(\sigma^2 r^2) + g_s^2(s)} \right]. \qquad (7.127)$$

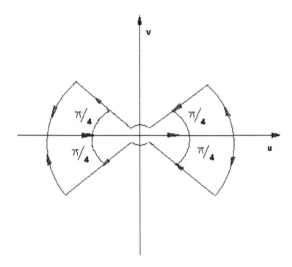

Fig. 7.8 The contour used to find the inverse of the Fourier transform of Eq. (7.100) for the jump length PDF $\lambda(x)$ given by Eq. (7.125).

Note that, in general, the inversion of the Laplace transform of Eq. (7.127), for a given $g(t)$, can be done; however it is difficult to obtain precise numerical results from it. In this case, it is more convenient to obtain the solutions from Eq. (7.94). Substituting Eq. (7.126) into Eq. (7.94) and applying the inverse Fourier transform yields

$$\rho_s(x,s) = \Psi_s(s) \left[ \delta(x) + \sum_{n=1}^{\infty} \frac{g_s^n(s) \exp[-\frac{x^2}{4n\sigma^2}]}{\sqrt{4\pi n\sigma^2}} \right]. \tag{7.128}$$

The $n$-moment (even $n$) related to Eq. (7.128) can be calculated exactly in Laplace space, and it is given by

$$\langle x^n \rangle_s = \frac{(4\sigma^2)^{\frac{n}{2}} \Gamma\left(\frac{1+n}{2}\right)}{\sqrt{\pi}} G_s(s) \sum_{j=1}^{\infty} j^{\frac{n}{2}} g_s^j(s). \tag{7.129}$$

Eq. (7.129) shows that the parameter $\sigma^2$ of the exponential jump length PDF does not change the scaling of the $n$-moment. In particular, the second moment can be written as follows:

$$\langle x^2 \rangle_s = 2\sigma^2 \frac{g_s(s)}{s(1 - g_s(s))}. \tag{7.130}$$

Note that the second moment (7.130) has the same result as the one given by the exponential jump length PDF for any waiting time PDF.

For $g_1(t)$ (Eq. (7.14)), we substitute Eq. (7.15) into Eq. (7.128) and apply the inverse Laplace transform, and the result is given by

$$\rho(x,t) = E_{\alpha,1}\left(-\lambda_\alpha t^\alpha\right)\delta\left(x\right)$$

$$+\sum_{j=1}^{\infty} \frac{(\lambda_\alpha t^\alpha)^j \exp\left(-\frac{x^2}{4j\sigma^2}\right) E_{\alpha,1}^{(j)}\left(-\lambda_\alpha t^\alpha\right)}{j!\sqrt{4\pi j\sigma^2}},$$  (7.131)

where $E_{\alpha,\beta}^{(j)}\left(y\right)$ (see Appendix 5.4.1) is given by

$$E_{\alpha,\beta}^{(j)}\left(y\right) = \frac{d^j E_{\alpha,\beta}\left(y\right)}{dy^j}.$$  (7.132)

The second moment, in this case, is given by

$$\left\langle x^2 \right\rangle = \frac{2\sigma^2}{\Gamma\left(1+\alpha\right)}t^\alpha,$$  (7.133)

which describes subdiffusive regime for all time intervals.

For $g_2(t)$ (Eq. (7.22)), the result for $\rho(x,t)$ is given by

$$\rho(x,t) = \frac{\Gamma\left(\beta, bt\right)}{\Gamma\left(\beta\right)}\delta\left(x\right)$$

$$+\sum_{j=1}^{\infty} \frac{\exp\left(-\frac{x^2}{4j\sigma^2}\right)}{\sqrt{4\pi j\sigma^2}}\left[\frac{\Gamma\left(\beta(1+j), bt\right)}{\Gamma\left(\beta(1+j)\right)} - \frac{\Gamma\left(\beta j, bt\right)}{\Gamma\left(\beta j\right)}\right],$$  (7.134)

where $\Gamma\left(\lambda, z\right)$ is the incomplete Gamma function [18]. Moreover, the $n$-moment can be obtained in the case of $g_2(t)$, and it has the following expression:

$$\left\langle x^n \right\rangle = \frac{(4\sigma^2)^{\frac{n}{2}}\Gamma\left(\frac{1+n}{2}\right)}{\sqrt{\pi}}\sum_{j=1}^{\infty} j^{\frac{n}{2}}\left[\frac{\Gamma\left(\beta(1+j), bt\right)}{\Gamma\left(\beta(1+j)\right)} - \frac{\Gamma\left(\beta j, bt\right)}{\Gamma\left(\beta j\right)}\right].$$  (7.135)

For $bt \ll 1$ the second moment yields

$$\left\langle x^2 \right\rangle \sim \frac{2\sigma^2 b^\beta}{\Gamma\left(1+\beta\right)}t^\beta,$$  (7.136)

whereas for $bt \gg 1$ it is given by

$$\left\langle x^2 \right\rangle \sim \frac{2\sigma^2 b}{\beta}t.$$  (7.137)

Eqs. (7.136) and (7.137) show that the second moment initially follows an anomalous regime (subdiffusive process for $0 < \beta < 1$ and superdiffusive process for $\beta > 1$) and it reaches a linear regime for large times. The linear regime given by Eq. (7.137) is due to the fact that the characteristic waiting time $T$ and the jump length variance $\Sigma^2$ are finite [1]. For $\beta = 1$, the $n$-moment (7.135) reduces to

$$\langle x^n \rangle = \frac{\left(4\sigma^2\right)^{\frac{n}{2}} \Gamma\left(\frac{1+n}{2}\right) e^{-bt}}{\sqrt{\pi}} \sum_{j=1}^{\infty} \frac{j^{\frac{n}{2}} (bt)^j}{j!}, \tag{7.138}$$

and the second moment reduces to the ordinary diffusion given by

$$\langle x^2 \rangle = 2\sigma^2 bt. \tag{7.139}$$

For $g_3(t)$ (Eq. (7.35)), a solution for $\rho(x,t)$ is given by

$$\rho(x,t) = A_1 \left( \frac{c_1}{\gamma_1} \exp\left(-\gamma_1 t\right) + \frac{c_2}{\gamma_2} \exp\left(-\gamma_2 t\right) \right) \delta(x)$$

$$+ \sum_{j=1}^{\infty} \frac{\exp\left(-\frac{x^2}{4j\sigma^2}\right) H(t,j)}{\sqrt{4\pi j\sigma^2}}, \tag{7.140}$$

where $H(t,n)$ is given by

$$H(t,n) = \left(\frac{A_1 c_1}{\gamma_1}\right)^n \left[ 1 - \frac{\Gamma(n, \gamma_1 t)}{\Gamma(n)} - \frac{A_1 c_1}{\gamma_1} \left(1 - \frac{\Gamma(1+n, \gamma_1 t)}{\Gamma(1+n)}\right) \right]$$

$$+ A_1^n \sum_{k1=1}^{n} \binom{n}{k1} c_2^{k1} c_1^{n-k1} t^n \left[ \frac{{}_1F_1(k1, 1+n, -\gamma_2 t)}{n!} \right.$$

$$+ (n - k1) \sum_{k2=0}^{\infty} \frac{(-\gamma_1 t)^{1+k2} (n - k1 + k2)! \, {}_1F_1(k1, 2+n+k2, -\gamma_2 t)}{(1+k2)! \, (n-k1)! \, \Gamma(2+n+k2)} \right]$$

$$- A_1^{1+n} \sum_{k1=1}^{1+n} \binom{1+n}{k1} c_2^{k1} c_1^{1+n-k1} t^{1+n} \left[ \frac{{}_1F_1(k1, 2+n, -\gamma_2 t)}{(1+n)!} + (1+n-k1) \right.$$

$$\left. \times \sum_{k2=0}^{\infty} \frac{(-\gamma_1 t)^{1+k2} (1+n-k1+k2)! \, {}_1F_1(k1, 3+n+k2, -\gamma_2 t)}{(1+k2)! \, (1+n-k1)! \, \Gamma(3+n+k2)} \right] \tag{7.141}$$

and ${}_1F_1(a,b,z)$ is the Kummer confluent hypergeometric function [15]. The $n$-moment related to $g_3(t)$ yields

$$\langle x^n \rangle = \frac{\left(4\sigma^2\right)^{\frac{n}{2}} \Gamma\left(\frac{1+n}{2}\right)}{\sqrt{\pi}} \sum_{j=1}^{\infty} j^{\frac{n}{2}} H(t,j). \tag{7.142}$$

The second moment can be simplified and it is described by

$$\langle x^2 \rangle = 2\sigma^2 \gamma_1 \gamma_2 \left\{ \frac{c_1 \gamma_2 + c_2 \gamma_1}{c_1 \gamma_2^2 + c_2 \gamma_1^2} t \right.$$

$$+ \frac{c_1 c_2 (\gamma_1 - \gamma_2)^2}{(c_1 \gamma_2^2 + c_2 \gamma_1^2)^2} \left[ 1 - \exp\left( -\frac{c_1 \gamma_2^2 + c_2 \gamma_1^2}{c_1 \gamma_2 + c_2 \gamma_1} t \right) \right] \left. \right\}. \tag{7.143}$$

For $t \ll 1$ and $t \gg 1$, the second moment of both limiting cases gives normal diffusive regimes. Moreover, for $\gamma_1 = \gamma_2$ the system recovers the normal diffusive process.

Detailed analyses of the diffusion behaviors of the uncoupled CTRW model have been investigated numerically [19] with the above solutions and the software Mathematica [15]. It should be noted that apart from these solutions there are other analytic expressions either in Laplace space or Fourier space, such as described in [20, 21], to deal with numerical work.

## 7.3 Integro-differential Fokker-Planck equation or integro-differential diffusion equation with external force

In fact, many physical systems are subject to the influence of an external force, such as a linear force described by the harmonic potential. We first consider the case of constant force $F(x) = F$, which may correspond, for instance, to the case of a constant electrical field acting on a charge particle. Now let us analyze the first two moments. From Eq. (6.33) we obtain

$$\frac{d \langle x(t) \rangle}{dt} = \int_0^t g(t - t_1) \frac{d \langle x(t_1) \rangle}{dt_1} dt_1$$

$$+ \frac{C}{k_B T} \frac{\partial}{\partial t} \int_0^t g(t - t_1) \int_{-\infty}^{\infty} F(x) \rho(x, t_1) dx dt_1 \tag{7.144}$$

and

$$\frac{d \langle x^2(t) \rangle}{dt} = \int_0^t g(t - t_1) \frac{d \langle x^2(t_1) \rangle}{dt_1} dt_1 + 2C \frac{\partial}{\partial t} \int_0^t g(t - t_1) \, dt_1$$

$$+ \frac{2C}{k_B T} \frac{\partial}{\partial t} \int_0^t g(t - t_1) \int_{-\infty}^{\infty} x F(x) \rho(x, t_1) dx dt_1. \tag{7.145}$$

Applying the Laplace transform to Eqs. (7.144) and (7.145) yields

$$\langle x(s) \rangle_{s,F} = \frac{F}{2k_B T} \langle x^2(s) \rangle_{s,F=0}, \tag{7.146}$$

with $\langle x(0)\rangle_F = \langle x^2(0)\rangle_{F=0} = 0$. Now applying the inverse Laplace transform to Eqs. (7.146) we obtain

$$\langle x(t)\rangle_F = \frac{F}{2k_BT} \langle x^2(t)\rangle_{F=0}. \tag{7.147}$$

The result (7.147) is very general since it is valid for generic waiting time PDF $g(t)$. Moreover, it is the well-known second Einstein relation which connects the first moment in the presence of constant force $F$ to the second moment without any external force. The relation (7.147) has also been verified experimentally in strongly disordered materials [22], in which there exists subdiffusive diffusion.

For a linear force $F(x) = -m\kappa^2 x$, with generic waiting time PDF $g(t)$, the first two moments can be solved by applying the Laplace transform to Eqs. (7.144) and (7.145), which are given by

$$\langle x(s)\rangle_s = \frac{\langle x(0)\rangle [1 - g_s(s)]}{s\left[1 - \left(1 - \frac{Cm\kappa^2}{k_BT}\right)g_s(s)\right]} \tag{7.148}$$

and

$$\langle x^2(s)\rangle_s = \frac{\langle x^2(0)\rangle [1 - g_s(s)]}{s\left[1 - \left(1 - \frac{2Cm\kappa^2}{k_BT}\right)g_s(s)\right]}$$

$$+ \frac{2Cg_s(s)}{s\left[1 - \left(1 - \frac{2Cm\kappa^2}{k_BT}\right)g_s(s)\right]}. \tag{7.149}$$

In the case of $g_1(t)$, the first two moments are given by

$$\langle x(t)\rangle = \langle x(0)\rangle E_{\alpha,1}(-\lambda_\alpha t^\alpha) \tag{7.150}$$

and

$$\langle x^2(t)\rangle = \frac{k_BT}{m\kappa^2} + \left(\langle x^2(0)\rangle - \frac{k_BT}{m\kappa^2}\right) E_{\alpha,1}(-2\lambda_\alpha t^\alpha). \tag{7.151}$$

It is found that the first two moments have the Mittag-Leffler decay. For $\alpha = 1$ all the above results reduce to those of normal diffusion described by the ordinary diffusion equation, and the Mittag-Leffler function decays like the exponential function. The thermal equilibrium is reached when $t \to \infty$, then we have $\langle x^2(\infty)\rangle = k_BT/m\kappa^2$. If the particle satisfies the special initial spatial condition as $\langle x(0)\rangle = 0$ and $\langle x^2(0)\rangle = k_BT/m\kappa^2$, then we have $\langle x(t)\rangle = 0$ and $\langle x^2(t)\rangle = k_BT/m\kappa^2$ for all the time. Therefore, average of displacement and its second moment are independent of time.

Exact solution for $\rho(x,t)$ can also be obtained by employing the method of separation of variables $\rho_n(x,t) = X_n(x)T_n(t)$; substituting it into Eq. (6.33) yields

$$\frac{\mathrm{d}T_n(t)}{\mathrm{d}t} - \int_0^t g(t-t_1)\frac{\mathrm{d}T_n(t_1)}{\mathrm{d}t_1}\mathrm{d}t_1 = -\mu_n\frac{\mathrm{d}}{\mathrm{d}t}\int_0^t g(t-t_1)T_n(t_1)\mathrm{d}t_1 \quad (7.152)$$

and

$$CL_{FP}X_n(x) = -\mu_n X_n(x), \qquad (7.153)$$

where $\mu_n$ are the eigenvalues. Then, the solution for $\rho(x,t)$ is given by the expansion of eigenfunctions

$$\rho(x,t\,|x',0) = e^{\frac{\overline{V}(x')}{2}-\frac{\overline{V}(x)}{2}}\sum_n \psi_n(x')\,\psi_n(x)\,T_n(t)\,, \qquad (7.154)$$

where $\overline{V}(x) = V(x)/k_BT$, $V(x)$ is the potential given by $F(x) = -\mathrm{d}V(x)/\mathrm{d}x$, and $\psi_n(x) = e^{\overline{V}(x)/2}X_n(x)$. We note that the eigenvalue equation of the operator $L_{FP}$, Eq. (7.153), is the same as the one of eigenvalue equation of the ordinary Fokker-Planck operator [10].

In the case of linear external force, the solution for the eigenvalue equation of the operator $L_{FP}$, Eq. (7.153), may be given in terms of the Hermite polynomials [10], and the solution for $\rho(x,t\,|x',0)$ is given by

$$\rho(x,t\,|x',0) = \frac{1}{\sqrt{2\pi x_{th}^2}}\sum_{n=0}^\infty \frac{T_n(t)}{2^n n!}H_n\left(\frac{\overline{x}}{\sqrt{2}}\right)H_n\left(\frac{\overline{x'}}{\sqrt{2}}\right)e^{-\frac{\overline{x}^2}{2}}, \quad (7.155)$$

where

$$\mu_n = nC/x_{th}^2, \qquad (7.156)$$

$\overline{x} = x/\sqrt{x_{th}^2}$, $H_n(y)$ denotes the Hermite polynomials and $x_{th}^2 = k_BT/(m\kappa^2)$ is the mean square position fluctuation. Besides, applying the Laplace transform to Eq. (7.152) yields

$$T_{sn}(s) = \frac{[1-g_s(s)]}{s[1-(1-\mu_n)g_s]}\,. \qquad (7.157)$$

It should be noted that the eigenvalue $\mu_n = n$ implies that

$$C = x_{th}^2, \qquad (7.158)$$

thus the first eigenvalue is zero which implies $T_0(t) = 1$. Then, the stationary solution can be found with $T_0(t) = 1$ and $\rho_{eq}(x) = \lim_{t\to\infty}\rho(x,t\,|x',0)$.

For the waiting time PDF $g_1(t)$, we obtain from Eq. (7.157) the following result:

$$T_n(t) = E_{\alpha,1}(-n\lambda_\alpha t^\alpha)\,. \qquad (7.159)$$

It is worth mentioning that this solution is the same as the one of the fractional diffusion equation (6.23). Then, the solution for $\rho(x, t \,|\, x', 0)$ is given by

$$\rho(x, t \,|\, x', 0)$$

$$= \frac{1}{\sqrt{2\pi x_{th}^2}} \sum_{n=0}^{\infty} \frac{E_{\alpha,1}(-n\lambda_\alpha t^\alpha)}{2^n n!} H_n\left(\frac{\overline{x}}{\sqrt{2}}\right) H_n\left(\frac{\overline{x'}}{\sqrt{2}}\right) e^{-\frac{\overline{x}^2}{2}}. \qquad (7.160)$$

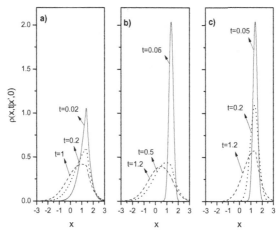

Fig. 7.9   Plots of PDF $\rho(x, t \,|\, x', 0)$ for $g_1(t)$, $g_2(t)$ and $g_3(t)$, at different time points. The initial value is chosen to be $x' = \sqrt{2}$ and $x_{th}^2 = 1$. a) The PDF for $g_1(t)$ with $\alpha = 0.5$ and $\lambda_\alpha = 1$. b) The PDF for $g_2(t)$ with $b = 0.5$ and $\beta = 0.5$. c) The PDF for $g_3(t)$ with a sum of two exponentials. The parameters have the following values: $c_0 = 0.5$, $\gamma_0 = 1$, $c_1 = 0.1$ and $\gamma_1 = 0.1$.

Fig. 7.9 shows the PDF $\rho(x, t \,|\, x', 0)$ as a function of $x$ coordinate at different time points for $g_1(t)$, $g_2(t)$ and $g_3(t)$. In Fig. 7.9.a, the PDF is plotted for $g_1(t)$, with $\alpha = 0.5$ and $\lambda_\alpha = 1$. The curves show cusps that are characteristic of anomalous diffusion behavior in the CTRW model. In Fig. 7.9.b, the PDF is plotted for $g_2(t)$, with $b = 0.5$ and $\beta = 0.5$. The curves show smooth shape due to the normal diffusion behavior. In Fig. 7.9.c, the PDF is plotted for $g_3(t)$, with a sum of two exponentials; in this case the intermediate curve is similar to the one plotted in Fig. 7.9.a, which deviates from a smooth shape. It is noted that all the positions of maxima of PDFs shift towards the origin with increase of time.

### 7.3.1  *Correlation function, mean square displacement, intermediate scattering function and dynamic structure factor*

In many cases the confinement of the atomic motions can be modeled by a linear force [23–25], and the model (6.33) in the presence of a harmonic trap may be used for the description of dynamics of a system (i.e., a tagged particle trapped in a harmonic potential, which describes the motion of all particles in the system), such as the internal motions of proteins. To this end we consider the correlation function, mean square displacement, intermediate scattering function and dynamic structure factor. Quantities, such as the intermediate scattering function and dynamic structure factor may be used to investigate systems submitted to neutron sources such as biological materials rich in hydrogen. For instance, neutron scattering technique has been employed to study myoglobin [26], superoxide dismutase [27] and $\alpha$-amylase [28].

The correlation function $C(t) = \langle x(t) x(0) \rangle$ is defined by the following relation:

$$C(t) = \int \int dx_0 dx x_0 x P(x, t \,|\, x_0, 0) \rho_{eq}(x_0) \,, \qquad (7.161)$$

where $P(x, t \,|\, x_0, 0)$ is the transition probability and $\rho_{eq}(x_0)$ is the equilibrium density given by

$$\rho_{eq}(x_0) = \frac{1}{\sqrt{2\pi x_{th}^2}} e^{-\frac{x_0^2}{2}} \,. \qquad (7.162)$$

The correlation function $C(t)$ may be obtained from Eq. (7.161) by substituting Eqs. (7.155) and (7.162) into Eq. (7.161) or from the operator $\int \int dx_0 dx x_0 x \rho_{eq}(x_0)$ applied to Eq. (6.33). The result is given by

$$C(t) = C(0) \left[ 1 - \int_0^t du\, g(u) \right], \qquad (7.163)$$

where $C(0)$ is the initial value of the correlation function. Note that the quantity on the right-hand side of Eq. (7.163) is the survival probability. Thus, the correlation function induced by the integro-differential diffusion equation (6.33) in the presence of the harmonic potential can be connected to the survival probability, i.e,

$$C(t) = C(0) \Psi(t) \,. \qquad (7.164)$$

The solution (7.163) is general, i.e., it is valid for generic waiting time PDF $g(t)$.

The form of $C(t)$ can also be evaluated by using the concept of the memory function and it can be described by

$$\frac{dC(t)}{dt} = -\frac{\partial}{\partial t} \int_0^t dt_1 g^*(t - t_1) C(t_1) . \tag{7.165}$$

Eq. (7.165) is obtained by applying the operator $\int \int dx_0 dx x_0 x \rho_{eq}(x_0)$ to Eq. (6.33). Note that the memory function associated with $C(t)$ is $g^*(t)$ of the integro-differential Fokker-Planck equation (6.35). Equation (7.165) generalizes the standard Ornstein-Uhlenbeck process and it contains a differential structure. In fact, the exponential form of $C(t)$ is obtained from Eq. (7.165) by using the exponential function for $g(t)$ which becomes an ordinary differential equation, in contrast to the standard one [23] given by

$$\frac{dC(t)}{dt} = -\int_0^t dt_1 \xi(t - t_1) C(t_1) , \tag{7.166}$$

where the exponential form of $C(t)$ is obtained by using $\xi(t) \propto \delta(t)$.

The mean square displacement measures the mean square value of the particle's displacement at time $t$ which is obtained from

$$\left\langle (x - x_0)^2 \right\rangle = \int dx_0 dx \, (x - x_0)^2 \, P(x, t | x_0, 0) \rho_{eq}(x_0). \tag{7.167}$$

Substituting Eqs. (7.155) and (7.162) into Eq. (7.167) yields

$$\left\langle (x - x_0)^2 \right\rangle = \frac{1}{2\pi x_{th}^2}$$

$$\times \sum_{n=0}^{\infty} \frac{T_n(t)}{2^n n!} \int dx_0 dx \, (x - x_0)^2 \, e^{-\frac{x^2}{2}} e^{-\frac{x_0^2}{2}} H_n\left(\frac{x}{\sqrt{2}}\right) H_n\left(\frac{x_0}{\sqrt{2}}\right). \tag{7.168}$$

Using the relation

$$H_n(x) = (-1)^n e^{x^2} \frac{d^n}{dx^n} \left(e^{-x^2}\right) \tag{7.169}$$

we obtain

$$\left\langle (x - x_0)^2 \right\rangle = \frac{1}{\pi x_{th}^2}$$

$$\times \sum_{n=0}^{\infty} \frac{T_n(t) x_{th}^{2n}}{n!} \int dx_0 dx \, (x^2 - xx_0) \left(\frac{d^n}{dx^n} e^{-\frac{x^2}{2x_{th}^2}}\right) \left(\frac{d^n}{dx_0^n} e^{-\frac{x_0^2}{2x_{th}^2}}\right)$$

$$= \frac{1}{\pi x_{th}^2} \left[ T_0(t) \int dx_0 dx x^2 e^{-\frac{x^2 + x_0^2}{2x_{th}^2}} - x_{th}^2 T_1(t) \left(\int dx_0 dx e^{-\frac{x^2}{2x_{th}^2}}\right)^2 \right]. \tag{7.170}$$

Solving the integrals of Eq. (7.170) yields

$$\left\langle (x - x_0)^2 \right\rangle = 2x_{th}^2 \left(1 - T_1(t)\right). \tag{7.171}$$

From Eq. (7.157) we obtain

$$T_1(t) = 1 - \int_0^t d\tau g(\tau) = \Psi(t). \tag{7.172}$$

We see that the function $T_1(t)$ results in the survival probability. The function $g(t)$ is a PDF, thus $\int_0^t d\tau g(\tau) \leq 1$; and, consequently, $T_1(t)$ is limited to $0 \leq T_1(t) \leq 1$. This means that the motions are confined in space due to the harmonic potential.

In neutron scattering experiments the atomic translational displacements, during the time interval $t$, are described by the intermediate scattering function, and it is defined by [23]

$$I_q(t) = \int dx_0 dx e^{iq(x-x_0)} \rho(x, t \,|\, x_0, 0) \rho_{eq}(x_0) \,, \tag{7.173}$$

where $q$ is the momentum transfer from the neutron to the sample in unit of $\hbar$. Substituting Eqs. (7.155) and (7.162) into Eq. (7.173) yields

$$I_q(t) = e^{-q^2 x_{th}^2} \sum_{n=0}^{\infty} \frac{\left(q^2 x_{th}^2\right)^n T_n(t)}{n!}. \tag{7.174}$$

Another important quantity measured in neutron scattering experiments is the dynamic structure factor, which is related to the intermediate scattering function by the Fourier transform:

$$S(q, \omega) = \frac{1}{2\pi} \int_{-\infty}^{\infty} dt e^{-i\omega t} I_q(t) \,, \tag{7.175}$$

where $\omega$ is the energy transfer from the neutron to the sample in unit of $\hbar$. Considering the function $T_n(t)$ is even in time, then Eq. (7.175) can be calculated using the Laplace transform, i.e,

$$S(q, \omega) = \frac{1}{\pi} \lim_{\varepsilon \to 0} \mathcal{R} \int_0^{\infty} dt e^{-(\varepsilon + i\omega)t} I_q(t) = \frac{1}{\pi} \lim_{\varepsilon \to 0} \mathcal{R} \left[I_{qs}(\varepsilon + i\omega)\right], \tag{7.176}$$

where $I_{sq}(s)$ is the Laplace transform of $I_q(t)$. Substituting Eq. (7.174) into Eq. (7.176) yields

$$S(q, \omega) = e^{-q^2 x_{th}^2}$$

$$\times \left\{ \delta(\omega) + \frac{1}{\pi} \sum_{n=1}^{\infty} \frac{\left(q^2 x_{th}^2\right)^n}{n!} \lim_{\varepsilon \to 0} \mathcal{R} \left[T_{ns}(\varepsilon + i\omega)\right] \right\}. \tag{7.177}$$

Note that the solutions (7.155), (7.171), (7.174) and (7.177) are general, i.e., they are valid for generic waiting time PDF $g(t)$.

Now the above quantities are investigated in details by employing the waiting time PDF $g_1(t)$, $g_2(t)$ and $g_3(t)$ given in section 7.2.1.

For $g_1(t)$ the correlation function is given by

$$C(t) = C(0)E_{\alpha,1}\left(-\lambda_\alpha t^\alpha\right). \tag{7.178}$$

This last result (7.178) is the same as the one obtained from the fractional diffusion equation (6.23) [23]; it is given by the Mittag-Leffler function which interpolates between the initial stretched exponential form and the power-law decay. For $\alpha = 1$, the standard Ornstein-Uhlenbeck process is recovered and the above quantity reduces to the exponential form

$$C(t) = C(0)e^{-\lambda t}. \tag{7.179}$$

The mean square displacement is obtained from Eq. (7.171) and it is given by

$$\left\langle (x - x_0)^2 \right\rangle = 2x_{th}^2 \left(1 - E_{\alpha,1}\left(-\lambda_\alpha t^\alpha\right)\right). \tag{7.180}$$

The mean square displacement (7.180) is also described by the Mittag-Leffler function which presents a power-law decay [29]. For $t \ll 1$ the mean square displacement follows a subdiffusive regime given by

$$\left\langle (x - x_0)^2 \right\rangle \sim \frac{2x_{th}^2 \lambda_\alpha}{\Gamma\left(1 - \alpha\right)} t^\alpha, \tag{7.181}$$

whereas for $t \gg 1$ it is given by

$$\left\langle (x - x_0)^2 \right\rangle \sim 2x_{th}^2 \left(1 - \frac{1}{\lambda_\alpha \Gamma\left(1 - \alpha\right) t^\alpha}\right). \tag{7.182}$$

The thermal equilibrium is reached when $t \to \infty$, then the mean square displacement converges to its plateau value $\left\langle (x - x_0)^2 \right\rangle = 2x_{th}^2$. It can be seen that the mean square displacement (7.182) converges slowly to its plateau value.

From Eq. (7.174) we obtain the following intermediate scattering function:

$$I_q\left(t\right) = e^{-q^2 x_{th}^2} \sum_{n=0}^{\infty} \frac{\left(q^2 x_{th}^2\right)^n E_{\alpha,1}\left(-n\lambda_\alpha t^\alpha\right)}{n!}. \tag{7.183}$$

Whereas, the dynamic structure factor is obtained from Eq. (7.46) and it has the form

$$S\left(q, \omega\right) = e^{-q^2 x_{th}^2} \tag{7.184}$$

$$\times \left\{ \delta\left(\omega\right) + \frac{1}{\pi}\sum_{n=1}^{\infty} \frac{\left(q^2 x_{th}^2\right)^n \tau_{\alpha,n} \sin\left(\frac{\alpha\pi}{2}\right)}{n! \left|\omega\tau_{\alpha,n}\right| \left[\left|\omega\tau_{\alpha,n}\right|^\alpha + 2\cos\left(\frac{\alpha\pi}{2}\right) + \left|\omega\tau_{\alpha,n}\right|^{-\alpha}\right]} \right\},$$

$$(7.185)$$

where the relaxation times $\tau_{\alpha,n}$ are given by

$$\tau_{\alpha,n} = \frac{1}{\left(n\lambda_\alpha\right)^{\frac{1}{\alpha}}}. \tag{7.186}$$

Note that the solutions (7.178), (7.183) and (7.185) have the same results as those obtained from the fractional diffusion equation (6.23) [23]. The standard Ornstein-Uhlenbeck (OU) process is recovered for $\alpha = 1$ and the above quantities reduce to

$$\left\langle \left(x - x_0\right)^2 \right\rangle = 2x_{th}^2 \left(1 - e^{-\lambda t}\right), \tag{7.187}$$

$$I_q\left(t\right) = e^{-q^2 x_{th}^2 \left(1 - e^{-\lambda t}\right)} \tag{7.188}$$

and

$$S\left(q,\omega\right) = e^{-q^2 x_{th}^2} \left\{ \delta\left(\omega\right) + \frac{1}{\pi}\sum_{n=1}^{\infty} \frac{\left(q^2 x_{th}^2\right)^n \tau_{1,n}}{n! \left[1 + \left(\omega\tau_{1,n}\right)^2\right]} \right\}. \tag{7.189}$$

For the waiting time PDF $g_2(t)$ the correlation function is described by

$$C(t) = C(0) \frac{\Gamma\left(\beta, bt\right)}{\Gamma\left(\beta\right)}, \tag{7.190}$$

where $\Gamma\left(a, z\right)$ is the incomplete Gamma function [18]. For $bt \ll 1$ the correlation function follows a power-law behavior given by

$$C(t) \sim C(0) \left[1 - \frac{(bt)^\beta}{\Gamma\left(1 + \beta\right)}\right], \tag{7.191}$$

whereas for $bt \gg 1$ it is given by

$$C(t) \sim C(0) \frac{(bt)^{\beta - 1} e^{-bt}}{\Gamma\left(\beta\right)}. \tag{7.192}$$

Fig. 7.10 shows the correlation function (7.190), and for comparison the exponential decay in the Brownian case.

The mean square displacement is given by

$$\left\langle \left(x - x_0\right)^2 \right\rangle = 2x_{th}^2 \left(1 - \frac{\Gamma\left(\beta, bt\right)}{\Gamma\left(\beta\right)}\right). \tag{7.193}$$

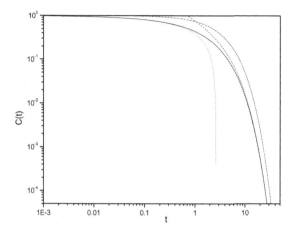

Fig. 7.10   Plots of the correlation function (7.190) (solid line) for the waiting time PDF $g_2(t)$ and its asymptotic limits (7.191) and (7.192) with $C(0) = 1$, $b = 0.3$ and $\beta = 0.5$. Whereas the dash-dotted line corresponds to the exponential decay in the Brownian case with $C(0) = 1$, $b = 0.3$ and $\beta = 1$.

For $bt \ll 1$ the mean square displacement also follows an anomalous regime (subdiffusion for $\beta < 1$ and superdiffusion for $\beta > 1$) given by

$$\left\langle (x - x_0)^2 \right\rangle \sim 2x_{th}^2 \frac{(bt)^\beta}{\Gamma(1+\beta)}, \tag{7.194}$$

whereas for $bt \gg 1$ it is given by

$$\left\langle (x - x_0)^2 \right\rangle \sim 2x_{th}^2 \left( 1 - \frac{(bt)^{\beta-1} e^{-bt}}{\Gamma(\beta)} \right). \tag{7.195}$$

Equation (7.195) shows that the mean square displacement of the second case converges more rapidly to its plateau value than the one of the first case due to the exponential decay.

The intermediate scattering function and the dynamic structure factor are given by

$$I_q(t) = e^{-q^2 x_{th}^2} \left[ 1 + q^2 x_{th}^2 \frac{\Gamma(\beta, bt)}{\Gamma(\beta)} + \sum_{n=2}^{\infty} \frac{(q^2 x_{th}^2)^n}{n!(1+n)} \right.$$

$$\times \sum_{l=0}^{\infty} \left( \frac{n}{1+n} \right)^l \sum_{j=0}^{l} \binom{l}{j} \left( \frac{1-n}{n} \right)^j \left( \frac{\Gamma(\beta(1+j), bt)}{\Gamma(\beta(1+j))} - \frac{\Gamma(\beta j, bt)}{\Gamma(\beta j)} \right) \right] \tag{7.196}$$

and

$$S(q, \omega) = e^{-q^2 x_{th}^2} \left\{ \delta(\omega) + \frac{1}{\pi} \right.$$

$$\left. \sum_{n=1}^{\infty} \frac{\left(q^2 x_{th}^2\right)^n n \left(\frac{b}{r}\right)^\beta \sin(\beta\theta)}{n! \, |\omega| \left[ \left( \cos(\beta\theta_2) - (1-n) \left(\frac{b}{r}\right)^\beta \right)^2 + \sin^2(\beta\theta) \right]} \right\}, \qquad (7.197)$$

where $r = \sqrt{b^2 + \omega^2}$ and $\theta = \arctan(\omega/b)$. The solutions (7.193)-(7.197) recover the standard OU process for $\beta = 1$.

For the waiting time PDF $g_3(t)$, it is given by a sum of two exponentials. The correlation function is described by

$$C(t) = C(0) A_1 \left( \frac{c_0}{\gamma_0} e^{-\gamma_0 t} + \frac{c_1}{\gamma_1} e^{-\gamma_1 t} \right). \qquad (7.198)$$

The mean square displacement is given by

$$\left\langle (x - x_0)^2 \right\rangle = 2 x_{th}^2 \left[ 1 - A_1 \left( \frac{c_0}{\gamma_0} e^{-\gamma_0 t} + \frac{c_1}{\gamma_1} e^{-\gamma_1 t} \right) \right]. \qquad (7.199)$$

For $t \ll 1$ the mean square displacement follows a normal regime given by

$$\left\langle (x - x_0)^2 \right\rangle \sim 2 x_{th}^2 A_1 (c_0 + c_1) t, \qquad (7.200)$$

whereas for $t \gg 1$ it is given by Eq. (7.199). Equation (7.199) shows that the mean square displacement converges exponentially to its plateau value.

The intermediate scattering function and the dynamic structure factor have the following forms:

$$I_q(t) = e^{-q^2 x_{th}^2} \left\{ 1 + q^2 x_{th}^2 A_1 \left( \frac{c_0}{\gamma_0} e^{-\gamma_0 t} + \frac{c_1}{\gamma_1} e^{-\gamma_1 t} \right) \right.$$

$$\left. + \sum_{n=2}^{\infty} \frac{\left(q^2 x_{th}^2\right)^n}{n!} \left[ \frac{(B_1 - D_1) e^{-D_1 t} - (B_1 - D_2) e^{-D_2 t}}{D_2 - D_1} \right] \right\} \qquad (7.201)$$

and

$$S(q, \omega) = e^{-q^2 x_{th}^2} \left\{ \delta(\omega) + \frac{1}{\pi} \right.$$

$$\left. \times \sum_{n=1}^{\infty} \frac{\left(q^2 x_{th}^2\right)^n n A_1 \left[ c_1 \gamma_0^2 + c_0 \gamma_1^2 + (c_0 + c_1) \omega^2 \right]}{n! \left\{ \left( n\gamma_0\gamma_1 - \omega^2 \right)^2 + \left[ \gamma_0 + \gamma_1 - (1-n) A_1 (c_0 + c_1) \right]^2 \omega^2 \right\}} \right\}, \qquad (7.202)$$

where

$$B_1 = \frac{c_1\gamma_0^2 + c_0\gamma_1^2}{c_1\gamma_0 + c_0\gamma_1}, \quad B_2 = B_1 + nA_1(c_0 + c_1), \quad (7.203)$$

$$B_3 = nA_1(c_1\gamma_0 + c_0\gamma_1), \quad D_1 = \frac{B_3}{D_2}, \quad D_2 = \frac{B_2 \pm \sqrt{B_2^2 - 4B_3}}{2}. \quad (7.204)$$

The solutions (7.198)-(7.202) also recover the standard OU process for $c_0 = 0$ or $c_1 = 0$.

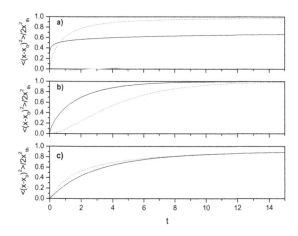

Fig. 7.11 Plots of the mean square displacement for $g_1(t)$, $g_2(t)$ and $g_3(t)$. a) The curves correspond to $g_1(t)$. The solid line corresponds to $\lambda_\alpha = 1$ and $\alpha = 0.2$ and the dotted line corresponds to $\lambda_\alpha = 1$ and $\alpha = 0.8$. b) The curves correspond to $g_2(t)$. The solid line corresponds to $b = 0.4$ and $\beta = 0.7$, and the dotted line corresponds to $b = 0.4$ and $\beta = 1.8$. c) The curves correspond to $g_3(t)$. The solid line corresponds to $\gamma_0 = 0.2$, $c_0 = 0.2$, $\gamma_1 = 0.5$ and $c_1 = 0.5$, and the dotted line corresponds to $\gamma_0 = 0.2$, $c_0 = 0.2$, $\gamma_1 = 0.9$ and $c_1 = 0.9$.

Fig. 7.11 shows the mean square displacement versus time for Eqs. (7.178), (7.193) and (7.199). All the curves increase with the time increase, however one of the curves shown in Fig. 7.11.b presents sigmoidal shape ($\beta = 1.8$) which is different from the other ones.

Fig. 7.12 shows the intermediate scattering function. All the curves decrease with the time increase and they converge to the same value $e^{-q^2 x_{th}^2}$. In Figs. 7.12.a and 7.12.b the curves show different features in contrast to the curves shown in Fig. 7.12.c.

Fig. 7.13 shows the dynamic structure factor for Eqs. (7.185), (7.197) and (7.202) without the contributions of $\delta(\omega)$. The spectra shown in Fig.

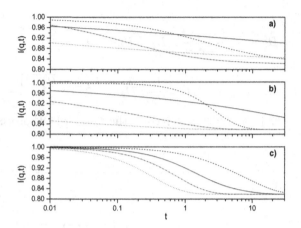

Fig. 7.12   Plots of the intermediate scattering function for $g_1(t)$, $g_2(t)$ and $g_3(t)$, with $q^2 x_{th}^2 = 0.2$. a) The curves correspond to $g_1(t)$. The solid, dotted, dashed and dash-dotted lines correspond to $\alpha = 0.2$ and $\lambda_\alpha = 0.5$; $\alpha = 0.2$ and $\lambda_\alpha = 2.5$; $\alpha = 0.6$ and $\lambda_\alpha = 0.5$; and $\alpha = 0.6$ and $\lambda_\alpha = 2.5$, respectively. b) The curves correspond to $g_2(t)$. The solid, dotted, dashed and dash-dotted lines correspond to $b = 0.005$ and $\beta = 0.2$; $b = 0.8$ and $\beta = 0.05$; $b = 0.5$ and $\beta = 1.5$; and $b = 0.5$ and $\beta = 0.2$, respectively. c) The curves correspond to $g_3(t)$. The solid, dotted, dashed and dash-dotted lines correspond to $\gamma_0 = 0.8$, $c_0 = 0.7$, $\gamma_1 = 0.2$ and $c_1 = 0.1$; $\gamma_0 = 2.5$, $c_0 = 0.7$, $\gamma_1 = 3.5$ and $c_1 = 0.5$; $\gamma_0 = 0.5$ $c_0 = 2.7$, $\gamma_1 = 0.1$ and $c_1 = 2.5$; and $\gamma_0 = 1.2$, $c_0 = 0.7$, $\gamma_1 = 1.2$ and $c_1 = 0.1$, respectively.

7.13.a present different behaviors for different values of $\alpha$. For $\alpha = 0.1$ the spectrum decays as a linear pattern in the interval of time considered; for $\alpha = 0.5$ it does not follows a linear pattern. Whereas, for $\alpha = 0.9$ the spectrum decays slowly initially (for $\omega < 0.2$), but it may be described by two linear regimes approximately. The spectra shown in Fig. 7.13.b follow a slow decay initially, then they decay almost linearly. The spectra shown in Fig. 7.13.c also present different behaviors for different values of $\gamma_0$, $c_0$, $\gamma_1$ and $c_1$. For small and intermediate values of the parameters, the spectra described by solid and dashed lines decay slowly initially, then they decay almost linearly. However, for small values of parameters ($\gamma_0 = c_0 = 0.02$) and intermediate values of parameters ($\gamma_1 = c_1 = 3$) the spectrum described by dotted line shows oscillatory pattern.

It is possible to construct a model in which the correlation function follows a given behavior. For instance, we want a model in which the correlation function has a power-law decay at intermediate times and asymptotic

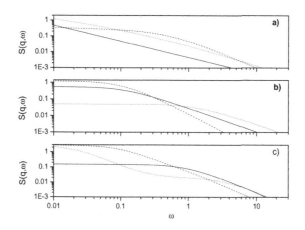

Fig. 7.13   Plots of the dynamic structure factor for $g_1(t)$, $g_2(t)$ and $g_3(t)$, with $q^2 x_{th}^2 = 0.5$. a) The curves correspond to $g_1(t)$. The solid, dotted and dashed lines correspond to $\alpha = 0.1$ and $\lambda_\alpha = 0.5$; $\alpha = 0.5$ and $\lambda_\alpha = 0.5$; and $\alpha = 0.9$ and $\lambda_\alpha = 0.5$, respectively. b) The curves correspond to $g_2(t)$. The solid, dotted and dashed lines correspond to $b = 0.1$ and $\beta = 0.5$; $b = 1.1$ and $\beta = 0.5$; and $b = 0.1$ and $\beta = 1.1$, respectively. c) The curves correspond to $g_3(t)$. The solid, dotted and dashed lines correspond to $\gamma_0 = 1$, $c_0 = 1$, $\gamma_1 = 0.9$ and $c_1 = 1.2$; $\gamma_0 = 0.02$, $c_0 = 0.02$, $\gamma_1 = 3$ and $c_1 = 3$; and $\gamma_0 = 0.5$, $c_0 = 0.3$, $\gamma_1 = 0.1$ and $c_1 = 0.5$, respectively.

exponential decay with a transition time $t_c$ described by [25, 30]

$$C(t) = [1 - \theta(t - t_c)] E_{\alpha,1}\left(-\left(\frac{t}{\tau_c}\right)^\alpha\right)$$

$$+ \theta(t - t_c) E_{\alpha,1}\left(-\left(\frac{t_c}{\tau_c}\right)^\alpha\right) e^{-\frac{t-t_c}{\tau_c}}, \qquad (7.205)$$

where $\theta(t)$ is the Heaviside function.

Eq. (7.205) describes a power-law decay of the correlation function at intermediate times for $t < t_c$ and asymptotic exponential decay for $t > t_c$. The correlation function given by Eq. (7.205) may be connected with the CTRW model, described in this section. To do so, Eq. (7.163) is rewritten as follows.

$$g(t) = -\frac{1}{C(0)} \frac{dC(t)}{dt}. \qquad (7.206)$$

Substituting Eq. (7.205) into Eq. (7.206) yields

$$g_4(t) = [1 - \theta(t - t_c)] \frac{t^{\alpha-1}}{\tau_c^\alpha} E_{\alpha,\alpha}\left(-\left(\frac{t}{\tau_c}\right)^\alpha\right)$$

$$+\frac{\theta\left(t-t_c\right)}{\tau_c}E_{\alpha,1}\left(-\left(\frac{t_c}{\tau_c}\right)^{\alpha}\right)e^{-\frac{t-t_c}{\tau_c}}$$

$$+\delta\left(t-t_c\right)\left[E_{\alpha,1}\left(-\left(\frac{t}{\tau_c}\right)^{\alpha}\right)-E_{\alpha,1}\left(-\left(\frac{t_c}{\tau_c}\right)^{\alpha}\right)e^{-\frac{t-t_c}{\tau_c}}\right],\qquad(7.207)$$

where $C(0)=1$. Eq. (7.207) is the $g(t)$ function for obtaining the correlation function (7.205).

One can observe that the integro-differential Fokker-Planck equation (6.33) has been developed in the structure of the CTRW model. It contains an enlarged structure which can also describe the dynamic behaviors of the ordinary and fractional diffusion equations. Equation (6.33), for force-free cases, can describe normal and subdiffusive regimes in the long-time limit, and it may also be used for improving the descriptions of ordinary and fractional diffusion equations, and be extended to describe other kinds of processes for short and intermediate times. Equation (6.33) has been investigated, in the presence of a harmonic trap, to obtain the mean square displacement, intermediate scattering function and dynamic structure factor. In fact, the generalized diffusion equation (6.33) can describe more general processes than the ordinary and fractional diffusion equations.

## 7.4   Generalized Rayleigh equation

The first two moments for velocity can be exactly obtained from Eq. (6.63) in Laplace space. The first moment for the velocity is calculated by

$$\frac{d}{dt}\langle v(t)\rangle$$

$$=\int_0^t dt_1 y\left(t-t_1\right)\frac{d\langle v(t_1)\rangle}{dt_1}-\tau_i\eta\frac{d}{dt}\int_0^t dt_1 q\left(t-t_1\right)\langle v(t_1)\rangle.\qquad(7.208)$$

Eq. (7.208) can be solved by using the Laplace transform and the result is

$$\langle v(s)\rangle_s=\frac{[1-g_s(s)]\langle v(0)\rangle}{s\left[1-\left(1-\tau_i\eta\right)g_s(s)\right]}.\qquad(7.209)$$

Eq. (7.209) is valid for generic waiting time PDF.

The second moment for the velocity is calculated by

$$\frac{d\langle v^2(t)\rangle}{dt}=\int_0^t dt_1 g\left(t-t_1\right)\frac{d\langle v^2(t_1)\rangle}{dt_1}$$

$$-2\tau_i\eta\frac{d}{dt}\int_0^t dt_1 g\left(t-t_1\right)\left[\langle v^2(t_1)\rangle-\frac{k_BT}{m}\right].\qquad(7.210)$$

Applying the Laplace transform yields

$$\langle v^2(s) \rangle_s = \frac{[1 - g_s(s)] \langle v^2(0) \rangle}{s[1 - (1 - 2\tau_i \eta) g_s(s)]}$$

$$+ 2\tau_i \eta \frac{k_B T}{m} \frac{g_s(s)}{s[1 - (1 - 2\tau_i \eta) g_s(s)]} . \tag{7.211}$$

It should be noted that for the case of waiting time PDF $g_1(t)$, the solution has the same result of the fractional Rayleigh equation. However, different behaviors for the time-dependence parts in the solution can be generated for other kinds of the waiting time PDF, such as the case of waiting time PDF $g_2(t)$. To obtain the PDF for velocity in time $t$, we employ the method of separation of variables $\rho_n(v,t) = X_n(v) T_n(t)$. Substituting $\rho_n(v,t)$ into Eq. (6.63) yields

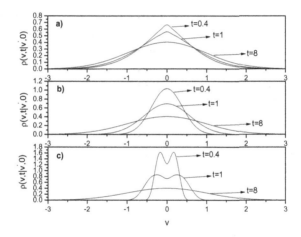

Fig. 7.14  Plots of the PDF $\rho(v, t \,|\, v', 0)$ for $g_2(t)$ at different time points $t = 0.4$, $1.0$ and $8.0$. The initial values are: $v' = 0$, $\eta \tau_i = 1$, $b = 0.2$ and $m/k_B T = 1$. a) for $\beta = 0.5$; b) for $\beta = 1$.; and c) for $\beta = 1.4$.

$$T_n(t) = \frac{1 - g_s(s)}{s[1 - (1 - \overline{n})g_s(s)]}, \quad n = 0, 1, 2, \dots , \tag{7.212}$$

and

$$\rho(v, t \,|\, v', 0) = \sqrt{\frac{m}{2\pi k_B T}} \sum_{n=0}^{\infty} \frac{T_n(t)}{2^n n!} H_n\left(\frac{\overline{v}}{\sqrt{2}}\right) H_n\left(\frac{\overline{v'}}{\sqrt{2}}\right) e^{-\frac{\overline{v}^2}{2}}, \tag{7.213}$$

where $T_n(0) = 1$, $\bar{v} = v\sqrt{m/(k_BT)}$, $\bar{n} = \eta\tau_i n$ and $H_n(y)$ denotes the Hermite polynomials.

For the waiting time PDF $g_2(t)$, the time-dependent parts $T_n(t)$ are given by

$$T_n(t) = 1 - \bar{n}b^\beta \int_0^t dt_1 E_{\beta,\beta}\left(-(\bar{n}-1)b^\beta t_1^\beta\right) t_1^{\beta-1}e^{-bt_1}. \tag{7.214}$$

When $\beta = 1$ the solution (7.213) reduces to the one of the usual Rayleigh equation [31]. Interesting behaviors of the PDF for velocity in different times are shown in Fig. 7.14. In Fig. 7.14.a, for $\beta < 1$, the PDF has one peak at earlier times and spreads out with the increase of time. Furthermore, it also shows up cusp at small and intermediate times. In Fig. 7.14.b, for $\beta = 1$, the PDF also has one peak, and the curves of the PDFs at different times are smoother than those of $\beta < 1$ at small and intermediate times. However, a peculiar result for the PDF comes out when $\beta > 1$, which is shown in Fig. 7.14.c. The PDF has two symmetric peaks at small and intermediate times, which corresponds to the tendency of a walker to continue moving in a direction over and above with select velocities.

## 7.5   Integro-differential Klein-Kramers equation

For the force-free case, the first two moments for displacement can be obtained exactly from Eq. (6.57) in Laplace space. The first moment of the displacement is calculated by

$$\frac{d\langle x(t)\rangle}{dt} = \tau_i \frac{\partial}{\partial t}\int_0^t dt_1 g^*(t-t_1)\langle v(t_1)\rangle, \tag{7.215}$$

where $g^*(t)$ is defined in Eq. (6.32). It should be noted that Eq. (7.215) does not obey the ordinary relation $d\langle x(t)\rangle/dt = \langle v(t)\rangle$, except for $y(t)$ given by an exponential function. Applying the Laplace transform to Eq. (7.215) and then substituting Eq. (7.209) into it yields

$$\langle x(s)\rangle_s = \frac{\tau_i g_s(s)\langle v(0)\rangle}{s\left[1-(1-\tau_i\eta)g_s(s)\right]}. \tag{7.216}$$

The second moment of the displacement is calculated by

$$\frac{d\langle x^2(t)\rangle}{dt} = 2\tau_i\frac{\partial}{\partial t}\int_0^t dt_1 g^*(t-t_1)\langle x(t_1)v(t_1)\rangle. \tag{7.217}$$

One can see that Eq. (7.217) also does not obey the ordinary relation $d\langle x^2(t)\rangle/dt = 2\langle x(t)v(t)\rangle$, either; except for $g(t)$ given by an exponential function. Therefore, the computation of the mean square displacement can not be made as the usual case. Applying the Laplace transform to Eq. (7.217) yields

$$\left\langle x^2(s) \right\rangle_s = \frac{\left\langle x^2(0) \right\rangle}{s} + \frac{2\tau_i g_s(s) \left\langle x(s)v(s) \right\rangle_s}{1 - g_s(s)} . \tag{7.218}$$

For $\langle x(t)v(t) \rangle$ one obtains

$$\frac{d\left\langle x(t)v(t) \right\rangle}{dt} = \int_0^t dt_1 g\left(t - t_1\right) \frac{d\left\langle x(t_1)v(t_1) \right\rangle}{dt_1}$$

$$+ \tau_i \frac{d}{dt} \int_0^t dt_1 \left[ g\left(t - t_1\right) \left\langle v^2(t_1) \right\rangle - \eta g\left(t - t_1\right) \left\langle x(t_1)v(t_1) \right\rangle \right] . \tag{7.219}$$

Applying the Laplace transform to Eq. (7.219) yields

$$\left\langle x(s)v(s) \right\rangle_s = \frac{\left\langle x(0)v(0) \right\rangle \left[1 - g_s(s)\right] + \tau_i s g_s(s) \left\langle v^2(s) \right\rangle_s}{s \left[1 - (1 - \tau_i \eta) g_s(s)\right]} . \tag{7.220}$$

Substituting Eq. (7.220) into Eq. (7.218) one obtains

$$\left\langle x^2(s) \right\rangle_s = \frac{2\tau_i^2 g_s^2(s) \left\langle v^2(s) \right\rangle_s}{\left[1 - g_s(s)\right] \left[1 - (1 - \tau_i \eta) g_s(s)\right]} , \tag{7.221}$$

with $\left\langle x^2(0) \right\rangle = \left\langle x(0)v(0) \right\rangle = 0$.

Thus, Eqs. (7.209), (7.211), (7.216) and (7.221) are the solutions in Laplace space for the first two moments. In general, the inversion of the Laplace transform of these quantities can be computed.

Eq. (6.57) can also be solved by using the method of separation of variables $\rho(x, v, t) = \overline{T}(t)\Xi(x, v)$ with

$$L_{KK}\Xi(x, v) = -\mu\Xi(x, v) \tag{7.222}$$

and

$$\frac{d\overline{T}(t)}{dt} = -\mu\tau_i \frac{\partial}{\partial t} \int_0^t dt_1 g^*\left(t - t_1\right) \overline{T}(t_1) , \tag{7.223}$$

where $\mu$ is the separation constant. It is noted that the solutions for $\Xi(x, v)$ are the same as those of eigenvalue equations for ordinary and fractional Klein-Kramers equations. However, Eq. (7.223) is different from time-dependence of ordinary and fractional Klein-Kramers equations. It can be solved by the Laplace transform as follows:

$$\overline{T}_s(s) = \overline{T}(0) \frac{1 - g_s(s)}{s \left[1 - (1 - \mu\tau_i) g_s(s)\right]} . \tag{7.224}$$

For the case of $g(t) = \delta(t - \Delta t)$, Eq. (6.57) reduces to the ordinary Klein-Kramers equation in the long-time limit by expanding $\rho(x, v, t - \Delta t)$ in power of $\Delta t$ up to first order. It is worth mentioning that Eq. (7.224) with $g_1(t)$ (Eq. (7.14)) gives the same time-dependence as the one of solution of the fractional Klein-Kramers equation [32]:

$$\overline{T}_\alpha(t) = \overline{T}(0)E_{\alpha,1}\left(-\mu_\alpha t^\alpha\right) . \qquad (7.225)$$

For $\alpha = 1$ one obtains the exponential function which is the same time-dependence of the solution of the ordinary Klein-Kramers equation.

In the case of the waiting time PDF $g_2(t)$ (Eq. (7.22)) yields

$$\overline{T}_2(t) = 1 - \mu\tau_i b^\beta \int_0^t dt_1 t_1^{\beta-1} \exp\left(-bt_1\right) E_{\beta,\beta}\left((1 - \mu\tau_i)(bt_1)^\beta\right) . \qquad (7.226)$$

It is noted that $\overline{T}_2(t)$ is very different from $\overline{T}_\alpha(t)$, and it results in a new diffusion behavior. In addition, $\overline{T}_2(t)$ recovers the limiting case of the ordinary Klein-Kramers equation when $\beta = 1$.

A solution of Eq. (6.57) may be obtained from the solution of the ordinary Klein-Kramers equation, i.e, it can be expressed in terms of the solution of the ordinary Klein-Kramers equation as follows. We consider the following transformation of variable:

$$\rho_s(x, v, s) = \frac{1 - g_s(s)}{\tau_i s g_s(s)}\rho_{s1}(x, v, s_1). \qquad (7.227)$$

Applying the Laplace transform to Eq. (6.57) yields

$$s\rho_s(x, v, s) - \rho(x, v, 0) = \frac{\tau_i s g_s(s)}{1 - g_s(s)}L_{KK}\rho_s(x, v, s). \qquad (7.228)$$

In order to obtain the ordinary Klein-Kramers equation the variable $s$ is transformed according to

$$s_1 = \frac{1 - g_s(s)}{\tau_i g_s(s)}. \qquad (7.229)$$

Substituting Eqs. 7.227 and 7.229 into Eq. 7.228 we obtain

$$s_1\rho_{s1}(x, v, s_1) - \rho(x, v, 0) = L_{KK}\rho_{s1}(x, v, s_1), \qquad (7.230)$$

where $\rho_{s1}(x, v, s_1, \eta_1)$ is the Laplace transform of the solution of the ordinary Klein-Kramers equation. The solution for $\rho(x, v, t)$ is obtained by inverting Eq. 7.227, i.e,

$$\rho(x, v, t) = \mathcal{L}^{-1}\left[\frac{1 - g_s(s)}{\tau_i s g_s(s)}\rho_{s1}\left(x, v, \frac{1 - g_s(s)}{\tau_i g_s(s)}\right)\right], \qquad (7.231)$$

where $\mathcal{L}^{-1}$ denotes the Laplace inversion. The solution of the ordinary Klein-Kramers equation for the force-free case is given by (see Section 4.3)

$$\rho_1(x, v, t \mid x', v', 0) = \frac{1}{2\pi\sqrt{A(t)}} \exp\left[ -\frac{A_{vv}(t)\,(x - x(t))^2}{2A(t)} \right.$$

$$\left. + \frac{A_{xv}(t)\,(x - x(t))\,(v - v(t))}{A(t)} - \frac{A_{xx}(t)\,(v - v(t))^2}{2A(t)} \right], \qquad (7.232)$$

where

$$A(t) = A_{xx}(t)A_{vv}(t) - A_{xv}^2(t), \qquad (7.233)$$

$$A_{xx}(t) = v_{th}^2 \left(2\eta t - 3 + 4\exp\left(-\eta t\right) - \exp\left(-2\eta t\right)\right)/\eta^2, \qquad (7.234)$$

$$A_{xv}(t) = v_{th}^2 \left(1 - \exp\left(-\eta t\right)\right)^2/\eta, \qquad (7.235)$$

$$A_{vv}(t) = v_{th}^2 \left(1 - \exp\left(-2\eta t\right)\right), \qquad (7.236)$$

$$v(t) = v' \exp\left(-\eta t\right), \qquad (7.237)$$

$$x(t) = x' + v' \left(1 - \exp\left(-\eta t\right)\right)/\eta, \qquad (7.238)$$

and $v_{th}^2 = k_B T/m$. Substitution of the Laplace transformation of the exponential waiting time PDF into Eq. (7.231) leads to $\rho_1(x, v, t)$. In Fig. 7.15 $\rho(x, v, t)$ is shown for the waiting time PDF $g_2(t)$ (Eq. (7.22)) obtained from Eq. (7.231) with the help of the numerical algorithm of the inverse Laplace transform [16](see Appendix 4.6.1). From Fig. 7.15, it is found that the PDF of the ordinary Klein-Kramers equation ($\beta = 1$) is more localized than the one of Eq. (6.57).

## 7.6   Generalized Klein-Kramers equation

Eq. (6.57) generalizes the ordinary and fractional Klein-Kramers equations (6.52). For the force-free case with a long-tailed power-law waiting time PDF Eq. (6.57) can describe subdiffusive behavior for large times. To describe superdiffusive regime, Barkai and Silbey [33] introduced the following equation:

$$\frac{\partial\rho(x, v, t)}{\partial t} = \left[ -v\frac{\partial}{\partial x} - \frac{F(x)}{m}\frac{\partial}{\partial v} + \gamma_{\alpha 0}D_t^{1-\alpha}L_R \right] \rho(x, v, t) . \qquad (7.239)$$

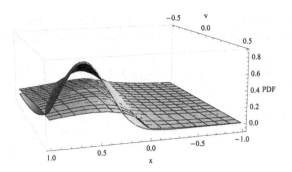

Fig. 7.15   Plots of the PDF $\rho(x, v, t \,|\, x', v', 0)$, using Eq. (7.231), $\rho_1(x, v, t \,|\, x', v', 0)$ and $g_2(t)$, for $t = 0.9$, in arbitrary units. The initial values are: $x' = 0$, $v' = 1$, $\eta = 0.5$, $b = 1/\tau_i = 0.5$ and $m/k_B T = 1$. The surface without mesh is obtained from the ordinary Klein–Kramers equation, whereas the surface with mesh corresponds to $\beta = 0.7$.

In Eq. (7.239), the fractional operator only acts on the part which contains the Rayleigh equation. An alternative model introduced by Metzler and Sokolov [34] is given by

$$\frac{\partial \rho(x, v, t)}{\partial t}$$

$$= \left[ -v \frac{\partial}{\partial x} - {_0}D_t^{1-\alpha} \varrho_\alpha \left( -\gamma \frac{\partial}{\partial v} v + \frac{F(x)}{m} \frac{\partial}{\partial v} - \kappa \frac{\partial^2}{\partial v^2} \right) \right] \rho(x, v, t). \quad (7.240)$$

The model (7.240) differs from the model (7.239) in the force field, where the action of the force enters through the non-local memory relation brought about by the fractional operator. In the absence of the force both models are equivalent, however, in the presence of the force they present different behaviors.

In order to obtain a generalized Klein-Kramers equation which can describe both the subdiffusive and superdiffusive regime, one takes the structure of Eq. (6.57), i.e. it can describe the same solutions as those obtained from the fractional Klein-Kramers equations (6.52), (7.239) and (7.240). Based on Eqs. (6.28) and (6.57) the PDF is described by [35]

$$\frac{\partial \rho(x, v, t)}{\partial t} - \int_0^t dt_1 g\left(t - t_1\right) \frac{\partial \rho(x, v, t_1)}{\partial t_1}$$

$$= -\tau_i \frac{\partial}{\partial t} \int_0^t dt_1 \left[ \bar{g}\left(t - t_1\right) v \frac{\partial}{\partial x} + \tilde{g}\left(t - t_1\right) \frac{F(x)}{m} \frac{\partial}{\partial v} \right] \rho(x, v, t_1)$$

$$+ \tau_i \eta \frac{\partial}{\partial t} \int_0^t dt_1 g\left(t - t_1\right) L_R \rho(x, v, t_1). \qquad (7.241)$$

Note that Eq. (7.241) recovers Eq. (6.57) for $\overline{g}(t) = \widetilde{g}(t) = g(t)$. Besides, Eq. (7.241) can also be written in a compact form given by

$$\frac{\partial \rho(x, v, t)}{\partial t} = \tau_i \frac{\partial}{\partial t} \int_0^t dt_1 \left\{ \eta g^* \left(t - t_1\right) L_R \rho(x, v, t_1) \right.$$

$$\left. - \left[ \overline{g}^* \left(t - t_1\right) v \frac{\partial}{\partial x} + \widetilde{g}^* \left(t - t_1\right) \frac{F(x)}{m} \frac{\partial}{\partial v} \right] \rho(x, v, t_1) \right\}, \qquad (7.242)$$

where $\overline{g}^*(t)$ and $\widetilde{g}^*(t)$ are defined through their Laplace transforms as follows:

$$\overline{g}_s^*\left(s\right) = \frac{\overline{g}_s(s)}{1 - g_s(s)} \qquad (7.243)$$

and

$$\widetilde{g}_s^*\left(s\right) = \frac{\widetilde{g}_s(s)}{1 - g_s(s)}. \qquad (7.244)$$

Eq. (7.242) can reduce to the fractional Klein-Kramers equations (7.239) and (7.240). To obtain Eq. (7.239) we need to take $g(t)$ given by Eq. (6.36) and $\widetilde{g}(t) = \overline{g}(t)$ given by the Riemann-Liouville fractional integral as follows.

$$\overline{g}(t) = \frac{1}{\lambda_\alpha \tau_i \Gamma(\nu)} \int_0^t dt_1 \frac{g(t_1)}{(t - t_1)^{1-\nu}}, \quad \nu = 1 - \alpha. \qquad (7.245)$$

The Laplace transform of Eq. (7.245) is given by

$$\overline{g}_s\left(s\right) = \frac{g_s(s)}{\lambda_\alpha \tau_i s^{1-\alpha}}. \qquad (7.246)$$

The Laplace transform of $g(t)$ described by Eq. (6.36) is given by $g_s(s) = \lambda_\alpha / \left(\lambda_\alpha + s^\alpha\right)$. Thus, we have

$$g_s^*\left(s\right) = \frac{\lambda_\alpha}{s^\alpha} \rightarrow g^*\left(t\right) = \frac{\lambda_\alpha}{\Gamma(\alpha) t^{1-\alpha}} \qquad (7.247)$$

and

$$\widetilde{g}_s^*\left(s\right) = \overline{g}_s^*\left(s\right) = \frac{\overline{g}_s\left(s\right)}{1 - g_s(s)} = \frac{1}{\tau_i s} \rightarrow \widetilde{g}^*\left(t\right) = \overline{g}^*\left(t\right) = \frac{1}{\tau_i}. \qquad (7.248)$$

Substituting the results (7.247) and (7.248) into Eq. (7.242) we obtain Eq. (7.239). To obtain Eq. (7.240) we need to take $\widetilde{g}(t) = g(t)$, $g(t)$ given by Eq. (6.36) and $\overline{g}(t)$ given by Eq. (7.245).

Despite the generalization made in Eq. (7.241), it maintains the same generalized Rayleigh equation (6.64) in velocity space. It can be checked by integrating Eq. (7.241) with respect to $x$.

### 7.6.1   *First two moments for velocity and displacement in the force-free case*

Exact solutions, in Laplace space, for the first two moments related to the model (7.241) with $\widetilde{g}(t) = \overline{g}(t)$ can be obtained, for the force-free case. The solutions for the first two moments for the velocity are the same as those of the generalized Rayleigh equation (6.64), and they are given by Eqs. (7.209) and (7.211). The first two moments of the displacement can be obtained following the method given in the previous section, which are given by

$$\langle x(s) \rangle_s = \frac{\tau_i \overline{g}_s(s) \langle v(0) \rangle}{s \left[ 1 - (1 - \tau_i \eta) g_s(s) \right]} \tag{7.249}$$

and

$$\langle x^2(s) \rangle_s = \frac{2\tau_i^2 \overline{g}_s^2(s) \langle v^2(s) \rangle_s}{\left[ 1 - g_s(s) \right] \left[ 1 - (1 - \tau_i \eta) g_s(s) \right]}, \tag{7.250}$$

with $\langle x^2(0) \rangle = \langle x(0)v(0) \rangle = 0$. The second moment of the velocity $\langle v^2(s) \rangle_s$ is given by Eq. (7.211).

Thus, Eqs. (7.209), (7.211), (7.249) and (7.250) are the solutions in Laplace space for the first two moments. In general, the inversion of the Laplace transform of these quantities can be done. Now the diffusion behaviors in two specific cases of waiting time PDF are studied.

*First case.* $g(t)$ and $\widetilde{g}(t) = \overline{g}(t)$ are described by Eqs. (7.14) and (7.245), respectively. For this case the solutions are the same as those obtained from the fractional Klein-Kramers equation (7.239) [33]. For the mean velocity yields

$$\langle v(t) \rangle = \langle v(0) \rangle E_{\alpha,1} \left( -\gamma_\alpha t^\alpha \right), \ \gamma_\alpha = \tau_i \eta \lambda_\alpha, \tag{7.251}$$

which exhibits a power-law decay described by the Mittag-Leffler function. For $\alpha = 1$, $\langle v(t) \rangle$ recovers the exponential decay.

The second moment $\langle v^2(t) \rangle$ is given by

$$\langle v^2(t) \rangle = \langle v^2(0) \rangle E_{\alpha,1} \left( -2\gamma_\alpha t^\alpha \right) + \frac{k_B T}{m} \left[ 1 - E_{\alpha,1} \left( -2\gamma_\alpha t^\alpha \right) \right], \tag{7.252}$$

and it also exhibits a power-law decay towards the thermal equilibrium value $\langle v^2(t \to \infty) \rangle \to k_B T/m$.

The first moment of the displacement $\langle x(t) \rangle$ yields

$$\langle x(t) \rangle = \langle v(0) \rangle t E_{\alpha,2} \left( -\gamma_\alpha t^\alpha \right), \tag{7.253}$$

which exhibits a net drift in the direction determined by the initial mean velocity $\langle v(0) \rangle$.

The mean square displacement $\langle x^2(t) \rangle$ is

$$\langle x^2(t) \rangle = 2 \frac{k_B T}{m} t^2 E_{\alpha,3} \left( -\gamma_\alpha t^\alpha \right) \tag{7.254}$$

with $\langle v^2(0) \rangle = k_B T / m$. For short times $\langle x^2(t) \rangle$ describes a ballistic diffusion

$$\langle x^2(t) \rangle \sim t^2, \tag{7.255}$$

whereas for long times $\langle x^2(t) \rangle$ exhibits superdiffusion

$$\langle x^2(t) \rangle \sim t^{2-\alpha} . \tag{7.256}$$

*Second case.* $\tilde{y}(t) = g(t)$ is described by Eq. (7.245) and the waiting time PDF $g(t)$ is described by an exponential function

$$g(t) = \mu e^{-\mu t} , \tag{7.257}$$

where $\mu > 0$. The mean velocity is

$$\langle v(t) \rangle = \langle v(0) \rangle \, e^{-\tau_i \eta \mu t} \tag{7.258}$$

which exhibits an exponential decay. The second moment $\langle v^2(t) \rangle$ is given by

$$\langle v^2(t) \rangle = \langle v^2(0) \rangle \, e^{-2\tau_i \eta \mu t} + \frac{k_B T}{m} \left[ 1 - e^{-2\tau_i \eta \mu t} \right] , \tag{7.259}$$

and it also exhibits an exponential decay towards the thermal equilibrium value $\langle v^2(t \to \infty) \rangle \to k_B T / m$.

The first moment of the displacement $\langle x(t) \rangle$ yields

$$\langle x(t) \rangle = \langle v(0) \rangle \frac{\mu}{\lambda_\alpha} t^{2-\alpha} E_{1,3-\alpha} \left( -\tau_i \eta \mu t \right) \tag{7.260}$$

which exhibits a net drift in the direction determined by the initial mean velocity $\langle v(0) \rangle$. For $\alpha = 1$, Eq. (7.260) reduces to the exponential form

$$\langle x(t) \rangle = \frac{\langle v(0) \rangle}{\tau_i \lambda_1 \eta} \left( 1 - e^{-\tau_i \eta \mu t} \right) . \tag{7.261}$$

The mean square displacement is obtained from Eqs. (7.250) and (7.257), and the result is

$$\langle x^2(t) \rangle = 2 \frac{k_B T}{m} \frac{\mu^2}{\lambda_\alpha^2} t^{4-2\alpha} E_{1,5-2\alpha} \left( -\tau_i \eta \mu t \right) . \tag{7.262}$$

The mean square displacement $\langle x^2(t) \rangle$ describes different superdiffusive behavior for short and long times. For short times it is given by

$$\langle x^2(t) \rangle \sim t^{4-2\alpha}, \tag{7.263}$$

whereas for long times it is given by

$$\langle x^2(t) \rangle \sim t^{3-2\alpha} . \tag{7.264}$$

It should be noted that the exponential waiting time PDF induces exponential decays for the velocity moments, but it induces power-law decays for the displacement moments. For $\alpha = 1$, the mean square displacement recovers the ordinary case, i.e, it describes a ballistic diffusion for short times and a normal diffusion for long-time limit. It is worth noticing that the above results are consistent with the corresponding waiting time PDFs. In the case of exponential waiting time PDF, the waiting time is shorter than that of the generalized Mittag-Leffler function (in statistical sense), thus the corresponding diffusion process is faster than the one of the first case.

### 7.6.2   *Solution of the generalized Klein-Kramers equation*

Eq. (7.241) can not be solved by the method of separation of variables due to the fact that the kernels $\overline{g}(t)$, $\widetilde{g}(t)$ and $g(t)$ are different. Besides, one can describe the solution of Eq. (7.241) in terms of the solution of the ordinary Klein-Kramers equation as follows. We consider the case $\widetilde{g}(t) = \overline{g}(t)$ and the following transformation of variable:

$$\rho_s(x, v, s, \eta) = \frac{1 - g_s(s)}{\tau_i s \overline{g}_s(s)} \rho_{s1}(x, v, s_1, \eta_1) . \tag{7.265}$$

From Eqs. (7.241) and (7.265) we obtain

$$\frac{1 - g_s(s)}{\tau_i \overline{g}_s(s)} \rho_{s1}(x, v, s_1, \eta_1) - \rho(x, v, 0)$$

$$= -\left[ v \frac{\partial}{\partial x} + \frac{F(x)}{m} \frac{\partial}{\partial v} \right] \rho_{s1}(x, v, s_1, \eta_1) + \frac{\eta g_s(s)}{\overline{g}_s(s)} L_R \rho_{s1}(x, v, s_1, \eta_1) . \tag{7.266}$$

In order to obtain the ordinary Klein-Kramers equation the variable $s$ and the coefficient $\eta$ are transformed according to

$$s_1 = \frac{1 - g_s(s)}{\tau_i \overline{g}_s(s)} \tag{7.267}$$

and

$$\eta_1 = \frac{\eta g_s(s)}{\overline{g}_s(s)}. \tag{7.268}$$

Substituting Eqs. (7.267) and (7.268) into Eq. (7.266) yields

$$s_1 \rho_{s1}(x, v, s_1, \eta_1) - \rho(x, v, 0) = -\left[ v\frac{\partial}{\partial x} + \frac{F(x)}{m}\frac{\partial}{\partial v} \right] \rho_{s1}(x, v, s_1, \eta_1)$$

$$+ \eta_1 L_R \rho_{s1}(x, v, s_1, \eta_1). \tag{7.269}$$

The solution $\rho(x, v, t, \eta)$ is obtained by the Laplace inversion of Eq. (7.265) which is described by

$$\rho(x, v, t, \eta) = \mathcal{L}^{-1} \left[ \frac{1 - g_s(s)}{\tau_1 s \overline{g}_s(s)} \rho_{s1}(x, v, s_1, \eta_1) \right], \tag{7.270}$$

where $\rho_{s1}(x, v, s_1, \eta_1)$ is the Laplace transform of the solution of the ordinary Klein-Kramers equation.

Substitution of the Laplace transformations of the exponential waiting time PDF and $\overline{g}(t)$ given by Eq. (7.245) into Eq. (7.270), with $\alpha = 1$, leads to $\rho_1(x, v, t)$. For $\overline{g}(t) = g(t)$, Eq. (7.270) reduces to the case of Eq. (6.57) [36]. When $g(t)$ is given by Eq. (6.36), Eq. (7.270) reduces to

$$\rho(x, v, t, \eta) = \mathcal{L}^{-1} \left[ \rho_{s1}(x, v, s, \gamma_\alpha s^{1-\alpha}) \right] \tag{7.271}$$

which has the same form as the one of the fractional Klein-Kramers equation introduced by Barkai and Silbey [33].

Eq. (6.57) is valid for generic waiting time PDF which is more general than ordinary and fractional Klein-Kramers equations. This means that it can reproduce the existing results from the ordinary and fractional diffusion equations. It has been shown the equivalence between the integro-differential equation (6.57) and Eq. (6.59) related to the CTRW model. Eqs. ((6.33), (6.57) and (6.64)) and their solutions provide a wider and systematic description of the CTRW model with generic waiting time PDF and external force. Moreover, analytical solutions can also be obtained by using the standard methods, and interesting dynamic behaviors of the PDFs both in phase space and in velocity space have been shown in various figures.

It is worthwhile to note the generalized Klein-Kramers equation (7.241) developed in this chapter is based on the structure of equations (6.28) and (6.57) which have been derived from the CTRW model. Eq. (7.241) contains an enlarged structure, and it can describe the dynamic behaviors of the ordinary and fractional Klein-Kramers equations (6.52) and the generalized Klein-Kramers equation (6.57), and it may extend to be used to describe other kinds of processes, which are not covered by the ordinary fractional equations. Moreover, Eq. (7.241) can reduce to the fractional Klein-Kramers equations (7.239) and (7.240). This means that the equation can describe subdiffusive and superdiffusive regimes of the previous equations and other kinds of processes. The extension of the formalism also modifies the ordinary mathematical relations, for instance, the ordinary relations $d\langle x(t)\rangle / dt = \langle v(t)\rangle$ and $d\langle x^2(t)\rangle / dt = 2\langle x(t)v(t)\rangle$ are replaced by relations (7.215) and (7.217).

## 7.7   Applications

The CTRW model was proved a useful tool for the description of nonequilibrium systems, from biophysics to geosciences. In fact, the CTRW has been used in a wide range of applications such as earthquake modelling [37, 38], random networks [39], self-organized criticality [40], electron tunneling [41], electron transport in nanocrystalline films [42], financial stock markets [43–48] and white and gray matters of a fixed rat brain [49]. For fractional and integro-differential Klein-Kramers equations (Eqs. (7.239) and (7.241)) which can describe superdiffusive regime they have been applied to the anomalous dynamics of cell migration of two migrating transformed renal epithelial MadinDarby canine kidney (MDCK-F) cell strains: wild-type (NHE+) and NHE-defficient (NHE-) cells [50, 51].

## 7.8   Appendices

### 7.8.1   *Wright function*

The Wright function, in the series representation, is defined by [2, 3]

$$W_{\lambda,\nu}(z) = \sum_{n=0}^{\infty} \frac{(z)^n}{n!\,\Gamma(\lambda n + \nu)} \ , \ z \in \mathbf{C} \ , \ \lambda > -1 \ , \ \nu > 0; \qquad (7.272)$$

it has the following integral representation:

$$W_{\lambda,\nu}(z) = \frac{1}{2\pi i} \int_{Ha} u^{-\nu} e^{u+zu^{-\lambda}} du \ , \ z \in \mathbf{C} \ , \ \lambda > -1 \ , \ \nu > 0, \qquad (7.273)$$

where $Ha$ denotes the Hankel path. The Wright function is a generalization of the exponential function, and it is also connected with the Mittag-Leffler function via the Laplace transform as follows:

$$\mathcal{L}\left[W_{\lambda,\nu}(z)\right] = \frac{1}{s} E_{(s)\lambda,\nu}\left(\frac{1}{s}\right) \ , \ \lambda > 0 \ , \ |s| > \nu > 0. \qquad (7.274)$$

The following relations are obtained from Eq. (7.272):

$$W_{0,1}(z) = e^z, \tag{7.275}$$

$$W_{-\frac{1}{2},\frac{1}{2}}(-z) = \frac{e^{-\frac{z^2}{4}}}{\sqrt{\pi}}, \tag{7.276}$$

$$\lambda z W_{\lambda,\lambda+\nu}(z) = W_{\lambda,\nu-1}(z) + (1 - \nu) W_{\lambda,\nu}(z), \tag{7.277}$$

$$\frac{dW_{\lambda,\nu}(z)}{dz} = W_{\lambda,\lambda+\nu}(z), \tag{7.278}$$

and

$$\lambda z \frac{dW_{\lambda,\nu}(z)}{dz} = W_{\lambda,\nu-1}(z) + (1 - \nu) W_{\lambda,\nu}(z). \tag{7.279}$$

# Bibliography

[1] R. Metzler and J. Klafter, *Phys. Rep.* **339**, 1 (2000).

[2] F. Mainardi, "Some Basic Problems in Continuum and Statistical Mechanics" in *Fractals and Fractional Calculus in Continuum Mechanics*, edited by A. Carpinteri and F. Mainardi (Springer, Wien, 1997), pp. 291-348.

[3] A. Erdélyi, W. Magnus, F. Oberhettinger and F. G. Tricomi, *Higher Transcendental Functions*, Bateman Project, Vol. 3 (McGraw-Hill, New York, 1955).

[4] M. E. Fisher and A. B. Kolomeisky, *PNAS* **98**, 7748 (2001).

[5] M. N. Berberan-Santos, *J. Math. Chem.* **38**, 165 (2005).

[6] K. S. Fa, *Phys. Rev. E* **82**, 012101 (2010).

[7] R. Metzler, J. Klafter and J. Jortner, *Proc. Natl. Acad. Sci. USA* **96**, 11085 (1999).

[8] J. Wlodarczyk and B. Kierdaszuk, *Biophys. J.* **85**, 589 (2003).

[9] I. Goychuk, *Phys. Rev. E* **80**, 046125 (2009).

[10] H. Risken, *The Fokker-Planck Equation: Methods of Solution and Applications*, 2rd ed. (Springer, Berlin, 1996).

[11] R. Gorenflo and F. Mainardi, "Fractional calculus: Integral and differential equations of fractional order" in *Fractals and Fractional Calculus in Continuum Mechanics*, edited by A. Carpinteri and F. Mainardi (Springer, Wien, 1997), pp. 223-276.

[12] M. Gitterman, *Phys. Rev. E* **62**, 6065 (2000).

[13] S. B. Yuste and K. Lindenberg, *Phys. Rev. E* **69**, 033101 (2004).

[14] E. Scalas, R. Gorenflo and F. Mainardi, *Phys. Rev. E* **69**, 011107 (2004).

[15] S. Wolfram, *The Mathematica*, 3rd ed. (Cambridge Univ. Press, Cambridge, 1996).

[16] H. Stehfest, *Commun. ACM* **13**, 47 (1970).

[17] K. S. Fa, *J. Physics A* **45**, 195002 (2012).

[18] I. S. Gradshteyn and I. M. Ryzhik, *Table of Integrals, Series and Products* (Academic Press, USA, 1965).

[19] K. S. Fa, *Phys. Rev. E* **89**, 052141 (2014).

[20] M. Dentz, A. Cortis, H. Scher, and B. Berkowitz, *Adv. Water Res.* **27**, 155 (2004).

[21] E. W. Montroll and H. Scher, *J. Stat. Phys.* **9**, 101 (1973).

[22] Q. Gu, E. A. Schiff, S. Grebner, F. Wang, and R. Schwarz, *Phys. Rev. Lett.* **76**, 3196 (1996).

[23] G.R. Kneller, *Phys. Chem. Chem. Phys.* **7**, 2641 (2005).

[24] S.H. Chen, C. Liao, F. Sciortino, P. Gallo and P. Tartaglia, *Phys. Rev. E* **59**, 6708 (1999).

[25] P. Calligari and D. Abergel, *J. Phys. Chem. B* **118**, 3823 (2014).

[26] W. Doster, S. Cusack and W. Petry, *Nature* **337** 754 (1989).

[27] A. Filabozzi, A. Deriu and C. Andreani, *Physica B* **226**, 56 (1996).

[28] J. Fitter, *Biophys. J.* **76**, 1034 (1999).

[29] R. Metzler and J. Klafter, *J. Phys. Chem. B* **104**, 3851 (2000).

[30] K. S. Fa, *Ann. Phys.* **353**, 179 (2015).

[31] N. G. van Kampen, *Stochastic Processes in Physics and Chemistry* (Elsevier, Hungary, 2004).

[32] R. Metzler and J. Klafter, *Phys. Rev. E* **61**, 6308 (2000).

[33] E. Barkai and R. J. Silbey, *J. Phys. Chem. B* **104**, 3866 (2000).

[34] R. Metzler and I. M. Sokolov, *Europhys. Lett.* **58**, 482 (2002).

[35] K. S. Fa, *J. Chem. Phys.* **137**, 234102 (2012).

[36] K. S. Fa and K. G. Wang, *Physica A* **391**, 3858 (2012).

[37] A. Helmstetter and D. Sornette, *Phys. Rev. E* **66**, 061104 (2002).

[38] Á. Corral, *Phys. Rev. Lett.* **97**, 178501 (2006).

[39] B. Berkowitz and H. Scher, *Phys. Rev. Lett.* **79**, 4038 (1997).

[40] M. Boguñá and Á. Corral, *Phys. Rev. Lett.* **78**, 4950 (1997).

[41] E. Gudowska-Nowak and K. Weron, *Phys. Rev. E* **65**, 011103 (2001).

[42] J. Nelson, *Phys. Rev. B* **59**, 15374 (1999).

[43] E. Scalas, R. Gorenflo and F. Mainardi, *Physica A* **284**, 376 (2000).

[44] F. Mainardi F, M. Raberto, R. Gorenflo and E. Scalas, *Physica A* **287**, 468 (2000).

[45] K. S. Fa, *Mod. Phys. Lett. B* **26**, 1250151 (2012).

[46] K. Kim and S. M. Yoon, *Fractals* **11**, 131 (2003).

[47] T. Kaizoji and M. Kaizoji, *Physica A* **336**, 563 (2004)

[48] J. Masoliver, M. Montero and G. H. Weiss, *Phys Rev. E* **67**, 021112 (2003).

[49] R. L. Magin, C. Ingo, L. Colon-Perez, W. Triplett and T. H. Mareci, *Microporous Mesoporous Mater* **178**, 39 (2013).

[50] P. Dieterich, R. Klages, R. Preuss and A. Schwab, *Proc. Natl. Acad. Sci. USA* **105**, 459 (2008).

[51] K. S. Fa and K. G. Wang, *J. Stat. Mech*, P09021 (2013).

Additional references for chapters 6-7 and fractional derivative approaches

[1] F. So and K. L. Liu, *Physica A* **331**, 378 (2004).

[2] C. W. Chow and K. L. Liu, *Physica A* **341**, 87 (2004).

[3] Y. P. Kalmykov, W. T. Coffey and S. V. Titov, *Phys Rev. E* **74**, 011105 (2006).

[4] Y. P. Kalmykov, W. T. Coffey and S. V. Titov, *Phys Rev. E* **75**, 031101 (2007).

[5] M. Magdziarz and A. Weron, *Phys Rev. E* **76**, 066708 (2007).

[6] Y. A. Rossikhin and M. V. Shitikova, *Appl. Mech. Rev.* **63**, 010801 (2010).

[7] Y. P. Kalmykov, W. T. Coffey and S. V. Titov, *Euro. Phys. J.* **222**, 1847 (2013).

[8] V. E. Tarasov, *Int. J. Mod. Phys.* **27**, 1330005 (2013).

[9] V. V. Uchaikin, *Fractional Derivatives for Physicists and Engineers* (Higher Education Press, Beijing and Springer-Verlag, Heidelberg, 2013).

[10] T. Sandev, A. Chechkin, H. Kantz and R. Metzler, *Fractional Calc. App. Anal.* **18**, 1006 (2015).

# Index